Applied Finite Mathematics

Mead

CENGAGE
Learning·

Australia • Brazil • Japan • Korea • Mexico • Singapore • Spain • United Kingdom • United States

Applied Finite Mathematics

Executive Editors:
 Maureen Staudt
 Michael Stranz

Senior Project Development Manager:
 Linda deStefano

Marketing Specialist:
 Courtney Sheldon

Senior Production/Manufacturing Manager:
 Donna M. Brown

PreMedia Manager:
 Joel Brennecke

Sr. Rights Acquisition Account Manager:
 Todd Osborne

Cover Image:
 Getty Images*

For product information and technology assistance, contact us at
Cengage Learning Customer & Sales Support, 1-800-354-9706

For permission to use material from this text or product,
submit all requests online at **cengage.com/permissions**
Further permissions questions can be emailed to
permissionrequest@cengage.com

This book contains select works from existing Cengage Learning resources and
was produced by Cengage Learning Custom Solutions for collegiate use. As such,
those adopting and/or contributing to this work are responsible for editorial
content accuracy, continuity and completeness.

Compilation © 2000 Cengage Learning

ISBN-13: 978-0-030-76821-7

ISBN-10: 0-030-76821-7

Cengage Learning
5191 Natorp Boulevard
Mason, Ohio 45040
USA

Cengage Learning is a leading provider of customized learning solutions with
office locations around the globe, including Singapore, the United Kingdom,
Australia, Mexico, Brazil, and Japan. Locate your local office at:
international.cengage.com/region.

Cengage Learning products are represented in Canada by Nelson Education, Ltd.
For your lifelong learning solutions, visit **www.cengage.com/custom.**
Visit our corporate website at **www.cengage.com.**

Printed in the United States of America

TABLE OF CONTENTS

CHAPTER ONE

SYSTEMS OF LINEAR EQUATIONS

1.1 SYSTEMS OF TWO LINEAR EQUATIONS IN TWO VARIABLES

OBJECTIVES:

1. *To solve systems of two linear equations in two variables by the method of elimination and the method of substitution.*

2. *To demonstrate by examples that a system of linear equations may have a unique solution, infinitely many solutions, or no solution.*

Consider the following example:

EXAMPLE 1: Find two integers whose sum is 54 and whose difference is 16.

Solution: If we let x, y denote the integers to be found, then the given conditions give the following relationship

$$x + y = 54 \tag{1}$$
$$x - y = 16 \tag{2}$$

This is an example of a linear system of two equations in two variables (unknowns). To solve the system, we add equations (1) and (2) to obtain

$$2x = 70$$

which is a linear equation in one variable and can be solved easily by multiplying both sides of the equation by $\frac{1}{2}$ to obtain

$$x = 35$$

Substituting x = 35 into (1), we have

$$35 + y = 54$$
$$\text{or} \quad y = 19$$

Therefore, 35 and 19 are the integers satisfying the given conditions.

The general form of a system of two linear equations in two variables (unknowns) may be written as

$$ax + by = k_1$$
$$cx + dy = k_2$$

where a, b, c, d are the __coefficients__ and k_1, k_2 are the constant terms. A solution of the system is an ordered pair (x,y) of numbers that satisfy both equations simultaneously. To solve a system of linear equations is to find all possible solutions of the system. Note in Example 1 it can be shown that the ordered pairs (x,y) = (35,19) is the only solution of the system. In this case we say the system has a __unique__ solution. It is also possible that a system of linear equations may have infinitely many solutions or no solution.

The system in Example 1 was solved by the *method of elimination*. The basic idea is to eliminate one of the variables from one of the equations by applying certain operations to be given below. The process is then reduced to solving one linear equation in one variable. This task can be accomplished by successive applications of the following operations:

1. Multiply an equation by a non-zero constant.

2. Add a multiple of one equation to the other.

We shall demonstrate this method in the following example.

EXAMPLE 2: Find the ages of John and his mother based on the following information:

(a) A year ago, twice the mother's age was ten times that of John.
(b) Four years from now nine times John's age will be the same as three times that of his mother.

Solution: We are asked to find the ages of John and his mother. We let x denote John's age and y his mother's age. It follows that a year ago John was x - 1 years old and his mother was y - 1. From information (a) we have the relation

$$2(y-1) = 10(x-1)$$

Since four years from now John will be x + 4 years old and his mother y + 4, information (b) gives us

$$9(x+4) = 3(y+4)$$

Therefore, we have the problem of solving the system

$$2(y-1) = 10(x-1) \tag{3}$$
$$9(x+4) = 3(y+4) \tag{4}$$

Equation (3) can be rewritten as

3

$$2y - 2 = 10x - 10$$

Substracting 10x and adding 2 to both sides, we have

$$-10x + 2y = -8 \tag{5}$$

Equation (4) can be rewritten as

$$9x - 3y = -24 \tag{6}$$

Thus, the system to solve is

$$-10x + 2y = -8 \tag{5}$$
$$9x - 3y = -24 \tag{6}$$

Apply operation 1 to both equations (5) and (6); that is, multiple Equation (5) by $\frac{1}{2}$ and Equation (6) by $\frac{1}{3}$ to obtain

$$-5x + y = -4 \tag{7}$$
$$3x - y = -8 \tag{8}$$

Adding Equation (7) and (8), we have

$$-2x = -12$$
$$x = 6 \quad \text{(John's age)}$$

Substitute x = 6 in (7) [or(8)] and we obtain

$$-5(6) + y = -4$$
$$y = 26 \quad \text{(His mother's age)}$$

Check: John is 6 years old and his mother is 26 years old. A year ago John was 5 and his mother was 25; hence

$$2 \cdot 25 = 50 = 10 \cdot 5$$

Thus Equation (3) is satisfied. Four years from now John will be 10 and his mother will be 30; hence

$$9 \cdot 10 = 90 = 3 \cdot 30$$

Thus, Equation (4) is satisfied and the solution is indeed correct.

SELF-TEST:

1. How many solutions can a system of two linear equations in two variables have?

    a. _____
    b. _____
    c. _____

Ans:

    a. a unique solution
    b. infinitely many solutions
    c. no solutions

2. The operations that are used in the method of elimination are:

    a. _____
    b. _____

Ans:

    a. multiply an equation by a non-zero constant
    b. add a multiple of one equation to another

3. Solve the system by the method of elimination

$$2x + 4y = 13$$
$$4x - 2y = 1$$

Solution: $(x,y) = \left(\frac{3}{2}, \frac{5}{2}\right)$

A system of two linear equations in two variables can also be solved by the *method of substitution*. That is, we express one variable in terms of the other in one of the equations and then substitute it into the other equation. We shall demonstrate this method by an example.

EXAMPLE 3: Solve the system

$$x - 2y = -4 \tag{9}$$
$$2x + 3y = 13 \tag{10}$$

Solution: Since the coefficient of x in equation (9) is 1, we express x in terms of y in equation (9). Thus we have

$$x = 2y - 4 \tag{11}$$

Substitute this value of x in (10) to obtain

$$2(2y-4) + 3y = 13$$
$$4y - 8 + 3y = 13$$
$$7y = 21$$
$$y = 3 \tag{12}$$

Substitute (12) back in (11), we have

$$x = 2$$

The solution of the system is $(x,y) = (2,3)$.

EXAMPLE 4: Rabbits and chickens are put together in a cage. The combined total is 12. If the total number of legs in the cage is 34, find the number of rabbits and the number of chickens in the cage.

Solution: Let x = the number of rabbits and y = the number of chickens. Knowing the fact that a rabbit has four legs and a chicken has only two, we derive the following system:

6

$$x + y = 12 \qquad\qquad (13)$$

$$4x + 2y = 34 \qquad\qquad (14)$$

Using the method of substitution, we solve for x in terms of y in (13) to obtain

$$x = 12 - y \qquad\qquad (15)$$

Substitute (15) into (14) to obtain

$$4(12-y) + 2y = 34$$
$$48 - 4y + 2y = 34$$
$$-2y = -14$$
$$y = 7$$

Substitute y = 7 in (15) to obtain

$$x = 5.$$

Therefore, there are 5 rabbits and 7 chickens in the cage.

Check: For equation (13),

$$5 + 7 = 12 \qquad \text{(true)}$$

For equation (14),

$$4 \cdot 5 + 2 \cdot 7 = 34 \quad \text{(true)}$$

SELF-TEST: Solve the following system by the method of substitution

$$x + 5y = 19$$
$$2x - 3y = -1$$

Solution: $(x,y) = (4,3)$

EXAMPLE 5: A fruit stand charges 40¢ for three apples and two pears and 70¢ for six apples and four pears. Can one find the cost of each apple and each pear at this fruit stand?

Solution: The unknowns are the costs for an apple and a pear. Thus, let x denote the cost in cents per apple and y denote the cost for one pear. The following system can be derived from the given information:

$$3x + 2y = 40 \qquad\qquad (16)$$
$$6x + 4y = 70 \qquad\qquad (17)$$

To use the method of elimination, we add -2 times (16) to (17) to obtain

$$0 = -10$$

which is impossible. Therefore, the system has no solution.

Notice that the system has no solution because the left side of (17) is twice that of (16) but the right side of (17) fails to be twice that of (16). We say this system is inconsistant.

A system of linear equations is said to be inconsistant if it has no solution.

SELF-TEST: Show that the following system has no solution.

$$4x - 3y = 2$$
$$12x - 9y = 5$$

We say that this system is _____.

Ans: inconsistant

If the sale of apples and pears in the last example is slightly altered as in the next example, we will have a system which has infinitely many solutions.

EXAMPLE 6: Suppose in Example 5, the cost for six apples and four pears is 80¢. What can one say about the solution of the resulting system?

Solution: In this case the system to solve is

$$3x + 2y = 40 \qquad\qquad (16)$$
$$6x + 4y = 80 \qquad\qquad (18)$$

Add -2 times (16) to (18) to obtain

$$0 = 0$$

Thus the system is reduced to only one linear equation (16) [or(18)]. Note here that (18) can be obtained by multiplying Equation (16) by 2. Therefore, every solution of (16) is a solution of (18) and vice versa. Equation (16) has infinitely many solutions which can be expressed as

$$x = \text{an arbitrary number}$$
$$y = \tfrac{1}{2}(40-3x)$$

For example, if we let x = 10, then (x,y) = (10,5) is a solution. That is, 10¢ per apple and 5¢ per pear is a possible price charged at the stand. Another solution is given by (x,y) = (4,14). On the other hand, (x,y) = (20,-10) is a solution of the system, but is discarded because -10 is not a possible price for the pear.

The solutions of the system are all the ordered pairs of numbers (x,y) such that y = $\tfrac{1}{2}$(40-3x). We may write the ordered pairs as (x,$\tfrac{1}{2}$(40-3x)). It is also convenient to express the solutions as a set

$$A = \{(x,y) \mid x \text{ any number}, y = \tfrac{1}{2}(40-3x)\}$$

It should be noticed here that we may also solve for x in terms of y in (16) to obtain

$$x = \tfrac{1}{3}(40-2y)$$

and the solution set is

$$B = \{(x,y) \mid y \text{ any number, } x = \tfrac{1}{3}(40-2y)\}$$

Note that the two sets A and B are equal.

SELF-TEST: Find all the possible solutions of the following system and express the solutions as a set (solution set).

$$5x - 7y = 3$$
$$-15x + 21y = -9$$

Ans: $\{(x,y) \mid x = \text{any}$
$y = \tfrac{1}{7}(5x-3)\}$ numbe.

From the above examples we see that a system of two linear equations in two variables may have

       1.   a unique solution,

or   2.   infinitely many solutions (this is the case if an identity such as 0 = 0 is obtained)

or   3.   no solution (this is the case when a statement such as 0 = -10 is obtained).

We remark that if a system has more than one solution then it must have infinitely many solutions.

The most commonly used method to solve a system of linear equations is the method of elimination. We summarize the procedure as follows:

STEP 1: Multiply each equation (or maybe just one equation

10

by a number to make the coefficient of one
of the x terms (or y terms) the negative of
the other.

STEP 2: Add the two resulting equations to eliminate
one of the variables x (or y).

STEP 3: Solve the linear equation in one variable x
(or y).

STEP 4: Substitute x (or y) into one of the equations in
the system to solve for y (or x).

NOTE: In Step 2, if both variables are eliminated, we have
one of the following two cases:

CASE 1: The identity 0 = 0 is obtained. In this case
the system is reduced to one linear equation
and there are infinitely many solutions.

CASE 2: The false statement 0 = a, where a is non-zero,
is obtained. In this case, the system has no
solution or is inconsistant.

We shall give one more example to demonstrate the
procedure.

EXAMPLE 7: Solve

$$2x + 4y = -14 \qquad\qquad (19)$$
$$3x - 5y = 34 \qquad\qquad (20)$$

Solution:

STEP 1: Multiply (19) by 3 and (20) by -2, to obtain

$$6x + 12y = -42 \qquad\qquad (21)$$
$$-6x + 10y = -68 \qquad\qquad (22)$$

11

Note that the coefficients of x in both resulting equations are 6 and -6 respectively.

STEP 2: Add (21) and (22) to eliminate x and obtain

$$22y = -110 \qquad\qquad (23)$$

STEP 3: Equation (23) is a linear equation in one variable y, solve for y:

$$y = -5$$

STEP 4: Substitute y = -5 into (21) to obtain

$$6x + 12(-5) = -42$$
$$x = 3$$

Therefore, the system has a unique solution $(x,y) = (3,-5)$.

SELF-TEST: Apply the above procedure to solve the system

1. $5x - 3y = -1$ (a)
    $2x + 4y = 10$ (b)

STEP 1: To have coefficients of x in both equations negative of each other we

    (1) multiple equation (a) by _____
    and obtain $-10x + 6y = 2$.   (c)

    (2) multiple equation (b) by _____
    and obtain $10x + 20y = 50$ (d)

STEP 2: To eliminate the variable x, add (c) and (d)
    to obtain _____.   (e)

STEP 3: Solve for y in (e). $y =$ _____.

12

STEP 4: Substitute y in (a) to obtain x = _____ .

Therefore, the solution of the system is
(x,y) = (,   )

Ans:

STEP 1. (1) -2
        (2) 5
STEP 2. 26y = 52
STEP 3. y = 2
STEP 4. x = 1
(x,y) = (1,2)

2. 3x + 5y = 2
    6x + 10y = 4

Ans: infinitely many
      solutions

3. 2x - 5y = 1
    -4x + 10y = 3

Ans: no solution.

EXERCISE 1.1:

Solve each of the following systems by the method of substitution

1. x = 3y - 5
    2x + 4y = 3

2. 2x + y = -3
    4x - 5y = 1

3. 5μ + 3V = 4
    μ - 2V = -7

4. 2r - s = -1
    3r + 4s = 6

5. 5x + 6y = 3
    3x - 4y = 2

6. 2x - 7y = 4
    5x + 2y = 3

Solve each of the following systems by the method of elimination.

13

7. $2x + 4y = 3$
$3x - 2y = 1$

8. $5x - 3y = 2$
$6x + 2y = -5$

9. $2x - 3y = 1$
$4x - 6y = 3$

10. $5x + 2y = 3$
$-15x - 6y = 4$

11. $-3x + 2y = 3$
$6x - 4y = -6$

12. $4x - 3y = 3$
$12x - 9y = 9$

13. $5\Lambda + 4s = 7$
$7\Lambda - 3s = -2$

14. $4\mu - 5V = -2$
$-3\mu + 4V = 3$

15. $0.2x + 3.4y = -2.3$
$4.1x - 2.7y = 0.45$

16. $3.4x - 11.3y = 4.7$
$-8.9x + 2.9y = -5.3$

17. $\frac{1}{2}x + \frac{3}{5}y = \frac{7}{4}$
$\frac{5}{8}x - \frac{4}{7}y = \frac{3}{4}$

18. $\frac{7}{3}x - \frac{5}{6}y = 4$
$\frac{2}{5}x + \frac{7}{6}y = 3$

Solve each of the following systems by either the method of substitution or by the method of elimination.

19. $x - 3y = 4$
$-3x + 2y = 7$

20. $5x + y = 4$
$-6x - 7y = 5$

21. $-2x + 3y = +4$
$5x - 2y = 1$

22. $7x - 6y = 7$
$3x + 8y = -2$

23. $0.5x - y = 0.3$
$-2.4x + 9.1y = 0.8$

24. $x + 1.52y = 3.4$
$7.5x - 6.34y = 0.785$

25. $\frac{5}{2}x + \frac{7}{2}y = \frac{11}{2}$
$\frac{8}{3}x - \frac{13}{3}y = \frac{25}{3}$

26. $-\frac{7}{9}x + \frac{1}{3}y = 4$
$\frac{1}{5}x + \frac{3}{10}y = -3$

27. $3x + y = 0$
$-2x - 3y = 0$

28. $\mu + V = 0$
$2\mu - V = 0$

29. Show that the following system has a unique solution if $ad - bc \neq 0$

$$ax + by = k_1$$
$$cx + dy = k_2$$

30. Show that if a, b, c are not zero, the following system has infinitely many solutions.

$$ax + by = k$$
$$cax + cby = ck$$

31. Show that if a, b, c are not zero and $k_1 \neq k_2$, then the following system has no solution.

$$ax + by = k_1$$
$$cax + cby = ck_2$$

32. For what value of a does the following system have infinitely many solutions?

$$2x - 3y = 4$$
$$-4x + 6y = a$$

33. The perimeter of a rectangle is 16 feet. If twice its length is four times its width, find the width and the length of the rectangle.

34. The perimeter of an isoceles triangle is 15 feet. If the unequal side is 3 feet less than the others, find the length of each side.

35. Find two integers if their sum is 19 and one is five greater than the other.

36. Find two numbers A and B if 5 times A is 4 less than 2 times B and B is 5 greater than A.

37. How old are Mary and her sister if seven years ago Mary was half as old as her sister and seven years from now Mary will be only three years younger?

38. John's father is 5 times as old as John. Five years from now his father will be only 3 times as old. How old will they be ten years from now?

39. Find the number of rabbits and chickens if the total number is 8 and total number of legs is 26.

40. The number of rabbits is 6 more than the chickens and the number of combined legs is 54. What are the numbers of rabbits and chickens?

41. The City of Clearlake has a population of 3000 more than that of the City of St. James. If we assume that each city will increase its population by 1000 in the next 10 years the population of Clearlake will then be twice that c St. James. Find the current population of each city.

42. Two dozen donuts and five cups of coffee are sold for $7.25. Four dozen donuts and 10 cups of coffee are sold for $10.25. Can you figure out the price per dozen donuts?

1.2 APPLICATIONS OF SYSTEMS OF TWO LINEAR EQUATIONS IN TWO VARIABLES

OBJECTIVES:

1. *To determine possible intersections of two lines.*
2. *To solve certain word problems involving a system of two equations in two variables.*

Intersection of lines:

An important application in solving a system of two equations in two variables is to find a point (or points) of intersection of two lines in a plane. Recall that a line in the plane can be represented as a linear equation in two variables.

$$ax + by = c.$$

Suppose two lines L_1 and L_2 are given as follows:

$$L_1: \quad ax + by = k_1 \qquad\qquad (1)$$
$$L_2: \quad cx + dy = k_2 \qquad\qquad (2)$$

To find a point of intersection of L_1 and L_2 one can solve the system of two linear equations (1) and (2).
If the system has a unique solution, the lines L_1 and L_2 have exactly one point of intersection (see Figure 1).

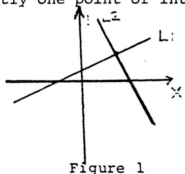

Figure 1

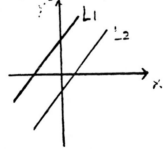

Figure 2

In case the system has no solution, then the lines L_1 and L_2 are said to be parallel, as shown in Figure 2.

Finally, if there are infinitely many solutions for the system, then the graphs of both equations are the same line (see Figure 3)

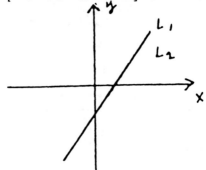

Figure 3

SELF-TEST:

1. To find a point of intersection of two lines one can solve a system of _____ linear equations in _____ variables.

2. If the system has only one solution, then the lines have _____ point(s) of intersection.

3. If the system has no solution, then the lines are _____.

4. If there are infinitely many solutions, the lines are _____.

ANS:

1. two, two
2. one
3. parallel
4. the same

EXAMPLE 1: Find the point of intersection of the two lines

18

L_1 and L_2 given by

$$L_1: \quad 2x - 5y = 4 \qquad\qquad (3)$$

$$L_2: \quad -3x + 2y = 5 \qquad\qquad (4)$$

Solution: We solve the system by the method of elimination to find a point of intersection. Multiply (3) by 3 and (4) by 2 to obtain

$$6x - 15y = 12 \qquad\qquad (5)$$
$$-6x + 4y = 10 \qquad\qquad (6)$$

Note that Equation (5) is obtained by multiplying Equation (3) by a non-zero constant . Hence, they represent the same line. Similarly, Equation (4) and (6) represent the same line. Now add (5) and (6) to obtain

$$-11y = 22$$
$$y = -2$$

Substituting $y = -2$ into Equation (3) we have

$$x = -3$$

Therefore, $(x,y) = (-3,-2)$ is the point of intersection of L_1 and L_2 as shown in Figure 4.

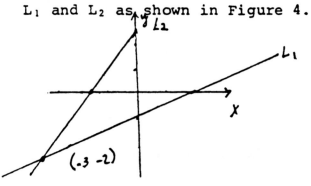

Figure 4

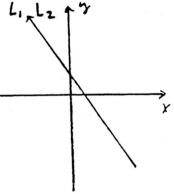

Figure 5

EXAMPLE 2: Given the two lines L_1 and L_2

19

$$L_1: \quad 5x + 2y = 3$$
$$L_2: \quad 10x + 4y = 6$$

Determine if they intersect at one point, are parallel, or represent the same line.

Solution: The equation representing L_2 is obtained by multiplying the equation representing L_1 by 2. This means they represent the same line as shown in Figure 5.

Suppose two non-vertical lines L_1 and L_2 are given as:

$$L_1: \quad y = \frac{-a}{b}x + \frac{k_1}{b}$$
$$L_2: \quad y = \frac{-c}{d}x + \frac{k_2}{d}$$

The terms $\frac{-a}{b}$, $\frac{-c}{d}$ are the *slopes* of L_1 and L_2 respectively. If the lines have the same slope, i.e, if $\frac{-a}{b} = \frac{-c}{d}$, then we compare the y-intercepts $\frac{k_1}{b}$ and $\frac{k_2}{d}$. If $\frac{k_1}{b} = \frac{k_2}{d}$, then they represent the same line. If they have distinct y-intercepts, then they are parallel.

EXAMPLE 3: Show that the lines L_1 and L_2 given below are parallel.

$$L_1: \quad 6x - 2y = 4$$
$$L_2: \quad -9x + 3y = 3$$

Solution: Solve for y in terms of x in each equation.

$$L_1: \quad y = 3x - 2$$
$$L_2: \quad y = 3x + 1$$

Both lines have the same slope 3 but different y-intercepts,

20

-2 and 1 respectively. Therefore, the lines are parallel
(no point of intersection).

SELF-TEST: Determine if each pair of lines given below
intersect at one point, infinitely many points, or no point.

1. y = 3x - 2
    y = 3x + 4

                                    Ans: parallel

2. 3x + 4y = 5
    -2x + 6y = 1

                                    Ans: intersect at
                                          one point.

3. 2x - 4y = 2
    3x - 6y = 3   Ans: coincide

4. 5x - 3y = 2
    10x - 6y = 1

                                    Ans: parallel

Business Related Applications:

    Some problems in the business world can be represented
as systems of linear equations. We illustrate one such problem
in the following example.

EXAMPLE 4: John invested $5,000 in two utility stocks.
Stock A pays dividends with an annual yield of 11% and Stock
B yields 12%. If John received $585 in dividends from both
stocks at the end of one year, how much did he invest in
each stock?

Solution: Since the unknown values are the amounts of money
invested in each stock, let x denote the amount of money
invested in Stock A and y denote the amount of money invested

21

in Stock B.   The total investment was known to be $5,000;
thus, we have the first equation

$$x + y = 5000 \qquad (7)$$

The dividend from Stock A with rate 11% after a year should
be 0.11x and that from Stock B with rate 12% should be 0.12y.
Based on the information that the total income from dividends
was $585, we derive the second equation

$$0.11x + 0.12y = 585 \qquad (8)$$

Solve the system of Equations (7) and (8) by the method
of substitution:

from (7), $x = 5000 - y$

$$\text{substitute in (8),} \quad 0.11(5000-y) + 0.12y = 585$$
$$550 - 0.11y + 0.12y = 585$$
$$-0.01y = -35$$
$$y = 3500$$

and $x = 5000 - 3500 = 1500$

Therefore, John invested $1500 in Stock A and $3500 in Stock B

EXAMPLE 5:   The Crown Food Company makes a food product named
Prince.   The food product contains two different ingredients,
sugar and flour.   It is required that for every two pounds of
sugar, three pounds of flour be added to the product.   If the
total weight of Prince is 500 pounds, how many pounds of sugar
and how many pounds of flour are in the product?

Solution: The unknowns are the weight of sugar and that of
flour in the product.   Let x denote the weight of sugar and
y denote the weight of flour mixed in Prince.   The total
weight is 500.   That is

22

$$x + y = 500 \qquad\qquad (9)$$

Knowing that for every 2 pounds of sugar there are 3 pounds of flour, we conclude that 3 times the weight of sugar must equal 2 times the total weight of flour. Therefore, we have the equation

$$3x = 2y \qquad\qquad (10)$$

The solution of the system with Equations (9) and (10) is

$$(x,y) = (200,300)$$

Hence, 200 pounds of sugar and 300 pounds of flour are mixed in the 500 pound food product Prince.

EXAMPLE 6: A manufacturing company produces two products A and B which require the labor of two skilled groups I and II. The number of working hours from each group to produce a unit of each product is given below:

	Group I	Group II
Product A	3 hrs.	4 hrs.
Product B	5 hrs.	6 hrs.

If the total number of available hours each week for Group I is 40 and that for Group II is 50, how many units of each product can be produced in one week?

Solution: Let x = the number of units produced for Product A each week. Let y = the number of units produced for Product B each week. The total number of hours required to produce x units of Product A and y units of Product B from Group I and Group II are given below:

	Group I	Group II
Product A	3x	4x
Product B	5y	6y

23

The number of available hours from Group I is 40 giving the equation

$$3x + 5y = 40 \qquad (11)$$

and that from Group II we have

$$4x + 6y = 50 \qquad (12)$$

Solving the system of Equations (11) and (12), we have

$$x = 5, \quad y = 5$$

Therefore, based on the available labor hours from the two groups, 5 units of each product, A or B, can be produced per week.

EXAMPLE 7: An oil company has two refineries A and B producing regular and premium gas. Their production per minute are as follows:

	Refinery A	Refinery B
Regular	40 barrels	20 barrels
Premium	30 barrels	40 barrels

The company received an order for 3000 barrels of regular and 5000 barrels of premium gas. How much time will the two refineries have to operate in order to produce the exact quantities ordered?

Solution: Let x denote the operating time for Refinery A and y denote that for Refinery B. It follows that the production amounts are:

	Refinery A	Refinery B
Regular	40x barrels	20y barrels
Premium	30x barrels	40y barrels

Since the order for regular is 3000 barrels, we have the relation

$$40x + 20y = 3000 \qquad\qquad (13)$$

and for the premium we have

$$30x + 40y = 5000 \qquad\qquad (14)$$

The solution obtained for the system consisting of (13) and (14) is

$$x = 20, \; y = 110$$

Hence, Refinery A must operate 20 minutes and Refinery B must operate 1 hour and 50 minutes to produce the desired amount.

EXAMPLE 8: Terry bought a tooth brush from a local drug store. She was overcharged by 72¢ due to a mistake. The two digits in the actual price of the tooth brush were reversed by the clerk. Find the actual price of the tooth brush bought by Terry.

Solution: This can be considered to be a digit problem. In order to solve it, we must recall that a two digit integer such as 23 is equal to 2 times 10 plus 3; that is, $23 = 2 \cdot 10 + 3$. Now, let x be the digit in the 10's position and y that in the unit position of the actual price. It follows that Terry should pay

$$(10x + y) \text{ cents}$$

for the tooth brush. The clerk at the store mistakenly reversed the two digits x, y; thus, Terry was charged

$$(10y + x) \text{ cents instead.}$$

The difference is

$$10y + x = 10x + y + 72 \tag{13}$$

We have only one equation in two variables which should have infinitely many solutions. Indeed, Equation (13) can be simiplified as

$$y - x = 8$$

and its solutions are

$$x \text{ arbitrary}$$
$$y = 8 + x$$

In this case, the reasonable solutions are

$$(x,y) = (0,8) \text{ and } (x,y) = (1,9)$$

since the digits can only be one of 0, 1,...,9. Therefore, the correct price for the tooth brush is either 8¢ or 19¢. It is very unlikely that the clerk would reverse 08 (8¢); hence 19¢ is the most possible correct price.

EXERCISE 1.2

In Problem 1-12, determine if each pair of lines are intersecting at a point, parallel or the same line.

1. $y = x$
   $x + y = 1$

2. $x - y = 3$
   $2x + y = 2$

3. $2x - y = -3$
   $2y - 4x = 5$

4. $x = 5y + 7$
   $y = 5x + 7$

5. $3x - 5y = 4$
   $4x + 8y = 2$

6. $5x - 3y = -1$
   $7x + 9y = 2$

26

7. $9x - 3y = 6$
   $3x - y = 2$

8. $2x + 4y = 10$
   $4x + 8y = 20$

9. $0.2x - 0.3y = -0.4$
   $1.7x + 3.1y = 2.5$

10. $-2.7x + 0.1y = -4.2$
    $-3.5x - 0.4y = 0.8$

11. $\frac{1}{2}x + \frac{1}{3}y = \frac{1}{5}$
    $\frac{1}{4}x + \frac{1}{7}y = \frac{1}{9}$

12. $\frac{1}{2}x - \frac{1}{6}y = \frac{1}{3}$
    $\frac{1}{5}x + \frac{3}{10}y = \frac{2}{5}$

13. Show that the two lines L_1 and L_2 have only one point of intersection if $ad - bc \neq 0$.

   L_1: $ax + by = k_1$
   L_2: $cx + dy = k_2$

14. What can you say about the intersection if $ad - bc = 0$ in Problem 13?

15. Mary received $5000 from her grandfather. She wants to invest the money to have an annual income of $510 to pay her tuition. If two investments are available yielding 10% and 11% per year respectively, how much should Mary invest in each to receive the exact amount of return?

16. $1200 is invested in two funds with an annual yield of 7% and 8% respectively. If an income of $89 is desired after a year, how much should be invested in each fund?

17. A food store makes two different kinds of food products, AII and BII. Food AII requires 5 minutes of machine time to process and costs $4 for ingredients while food BII requires 2 minutes and costs $2. How many of each kind can be made each day so that the store spends exactly $440 on the ingredients and the machine runs exactly 8 hours a day?

18. An oil company produces regular and unleaded gas from its two refineries. In one hour Refinery A can produce 5000 barrels of regular and 3000 barrels of unleaded, while Refinery B produces 2000 barrels of regular and 6000 of unleaded. How much time should each refinery be in operation in order to produce altogether 39,000 barrels of regular and 57,000 barrels of unleaded gas?

19. A shoe store is intending to buy and store two different sizes of shoes; size S and size L, all come with boxes. The size S shoe costs $20 per pair and size L $30. The box having size S is 1 square foot and that with size L is 2 square feet. The store intends to spend a total of $9000 to order the shoes and to fill the available 500 squre feet storage space. How many pairs of each size should be ordered?

20. Three dozen pencils and two dozen pens cost a total of $5.40. Ten dozen pencils costs $1.20 less than four dozen pens. How much does each dozen of pencils and each dozen of pens cost? (.60¢, $1.80)

21. Five hot dogs and three hamburgers cost a total of $6.20. Two hot dogs cost 50¢ more than a hamburger. How much does each hot dog and each hamburger cost?

22. The City of Tara is planning to make a float to enter the annual rose parade. The float needs 100 dozen roses and carnations for decoration. For every three dozen carnations, four dozen roses must be added in the decoration. The city spent $14,750 in total for buying the flowers. How many dozen roses and carnations has the City of Tara bought for the float if roses and carnations cost the same?

23. Michael paid $5.90 for his birthday party to buy four ice creams and 6 banana splits. Two ice creams are 5¢ cheaper than one banana split. How much does each

banana split and each ice cream cost?

24. It costs the theatre owner $8,000 to put out a rock concert. The owner anticipated that 700 adult and 300 child's tickets would be sold. If the owner would like to make $2,800 profit from the concert, how much should be charged for each adult and each child's ticket if the sum must be $5.00?

25. The charge for admission to a local amusement park is $2.25 for adults and 75¢ for children. There are twice as many children as adults admitted to the park and the total receipt for admission is $375. Find the number of adults and the number of children who visited the park.

26. Pat received $1.95 change from the clerk at the super market. The change is in quarters and dimes. If three times the number of quarters is one more than 2 times that of dimes, how many quarters and dimes did Pat receive?

27. Three pounds of coffee and two pounds of sugar cost $ 1.50 less than 4 pounds of coffee. How much does each pound of coffee and sugar cost, if coffee is four times as expensive as sugar?

28. Dora has 200 pounds of cloth to wash at a laundry mat. There are two sizes of washing machines in the store, large and small. A large size machine will wash 35 pounds of cloth and charge 3 quarters per load and the small size machine takes 20 pounds of cloth and costs 2 quarters per load. If Dora spent $4.50 on washing her cloth, for how many loads did she use the large size machine and for how many the small size?

29. A positive integer has two digits. If the integer becomes 26 greater after the digits are reversed and one digit is 4 less than the other, find the two digits.

30. Cora was over charged $2.72 at the checkout counter of the supermarket. Instead of charging her for 5 cans of tuna and 3 cans of shrimp, she was charged $13.12 for 5 cans of shrimp and 3 cans of tuna. How much did each can of tuna and each can of shrimp cost?

1.3 GENERAL SYSTEM OF LINEAR EQUATIONS AND METHOD OF ELIMINATION

OBJECTIVES:

1. *To solve a system of linear equations by the Gaussain elimination methods.*
2. *To solve applied problems involving systems of linear equations.*
3. *To introduce systems of linear homogeneous equations.*

A general system of m equations in n variables can be expressed as follows:

$$a_{11} x_1 + a_{12} x_2 + \ldots + a_{1n}x_n = b_1$$

$$a_{21} x_1 + a_{22} x_2 + \ldots + a_{2n}x_n = b_2$$

$$\vdots \qquad \vdots \qquad \qquad \vdots \qquad \vdots$$

$$a_{m1} x_1 + a_{m2} x_2 + \ldots + a_{mn}x_n = b_m$$

where x_1, x_2, ..., x_n are the variables, a_{11}, a_{12}, ..., a_{mn} are the coefficients of the variables and b_1, b_2, ..., b_n are the constant terms of each equation. If m = n = 2, we then have a system of two equations in two variables which has been discussed in detail in Sections 1.1 and 1.2. A solution for the system is an ordered n-tuple $(x_1, x_2, \ldots x_n)$ of numbers which satisfies all the m equations simultaneously. To solve a system is to find all solutions of the system if they exist. The most commonly used method to solve a system of linear equations is the Gaussain elimination method. The basic idea of the method used to solve a general system is basically the same as that for a system of two equations in two variables. However, there will possibly be more variables involved.

Two systems of linear equations are said to be *equivalent* if they have exactly the same solutions. For example, the following two systems are equivalent:

$$\text{I} \qquad \begin{aligned} 2x - 3y &= -4 \\ x + y &= 3 \end{aligned}$$

$$\text{II} \qquad \begin{aligned} 2x &= 2 \\ 5x - 3y &= -1 \end{aligned}$$

They are equivalent because both systems have the unique solution $(x,y) = (1,2)$. It is clear that System II is easier to solve than System I. Since the first equation in System II has only one variable, x, its value, 1, can be readily obtained by multiplying both sides of the equation by $\frac{1}{2}$. Substitute the value $x = 1$ in the second equation of System II, we can easily obtain the value $y = 2$. Our aim is to apply the elimination method to change a given system into an equivalent one whose solution (or solutions) can be easily obtained. Since they are equivalent, the solution set of the resulting system is also the one for the given system.

The operations allowed to change a system into its equivalence are the following:

(1) Multiplying an equation by a non-zero constant.

(2) Adding a multiple of one equation to another.

(3) Interchanging two equations.

We have seen the elimination method used to solve a system of two equations in two variables. In that special case, only one variable needed to be eliminated. For the general case, we shall try to eliminate as many variables

32

as possible and hopefully a linear equation in one variable can be reached. The procedure can be briefly described as follows:

STEP 1. Choose a variable from one of the equations. (If possible, choose the one with coefficient 1) Interchange that equation with the first equation if necessary (for convenience only). Then eliminate the same variable from the remaining equations by applying operations (2) and (3) respectively.

Note that after STEP 1, we obtain an equivalent system in which all but the first equation has one and the same variable missing.

STEP 2. Repeat the same process for the system with the first equation excluded until no variable can be eliminated.

STEP 3. Solve the last equation in the system obtained from the end of STEP 2. Substitute the value (or values) obtained for the variable (or variables) back to the next to the last equation which has at least one more variable than the last equation just solved and repeat the process until all variables are solved.

We shall demonstrate the procedure by solving the following system.

EXAMPLE 1: Apply the method of elimination to solve the system

$$2x - 3y + 2z = 9 \qquad (1)$$
$$x + 2y - z = -3 \qquad (2)$$
$$3x - y + 5z = 14 \qquad (3)$$

<u>Solution:</u>

STEP 1. In Equation (2), the coefficient of x is 1. We thus use Equation (2) to eliminate the variable x from Equations (1) and (3). First, we interchange (1) and (2) and rewrite the systems as follows:

$$x + 2y - z = -3 \qquad (2)$$
$$2x - 3y + 2z = 9 \qquad (1)$$
$$3x - y + 5z = 14 \qquad (3)$$

To eliminate 2x from (1), add -2 times Equation (2) to Equation (1):

$$-7y + 4z = 15 \qquad (4)$$

To eliminate 3x from Equation (3), add -3 times Equation (2) to Equation (3):

$$-7y + 8z = 23 \qquad (5)$$

We thus have an equivalent system

$$x + 2y - z = -3$$
$$-7y + 4z = 15$$
$$-7y + 8z = 23$$

Notice that the variable x is missing from Equations (4) and (5).

STEP 2. Now consider the system

$$-7y + 4z = 15 \qquad (4)$$
$$-7y + 8z = 23 \qquad (5)$$

We shall eliminate the variable y from Equation (

34

by adding -1 times Equation (4) to Equation (5):

$$4z = 8 \tag{6}$$

There are no more variables to be eliminated so we move to the next step. Note here we have obtained the equivalent system

$$x + 2y - z = -3 \tag{2}$$
$$-7y + 4z = 15 \tag{4}$$
$$4z = 8 \tag{6}$$

STEP 3. Solve the last equation (6) for z:

$$z = 2$$

Substitute z = 2 in (4) to obtain the value for y:

$$-7y + 4 \cdot 2 = 15$$
$$y = -1$$

Substitute both z = 2, y = -1 in Equation (2) to obtain the value of x:

$$x + 2 \cdot (-1) - 2 = -3$$
$$x = 1$$

Hence, (x,y,z) = (1,-1,2) is the solution of the given system.

SELF-TEST:

1. Two systems of linear equation are said to be equivalent if they have _____.

<u>Ans:</u> the same solutions

2. The operations allowed to change a system to its
   equivalence are:

   a. _____
   b. _____
   c. _____

The system in Example 1 has a unique solution. It is possible, as in the case of systems of two equations in two variables, that a general system of linear equations may have infinitely many solutions or no solutions (inconsistant system).

EXAMPLE 2: Solve

$$x_1 + 2x_2 - x_3 + 4x_4 = 11 \tag{7}$$
$$2x_1 + 5x_2 + x_3 + 6x_4 = 13 \tag{8}$$
$$3x_1 + 7x_2 + 2x_3 + 14x_4 = 22 \tag{9}$$

Solution:

STEP 1. Use Equation (7) to eliminate x_1 from Equation (8) and Equation (9). Add -2 times Equation (7) to Equation (8):

$$x_2 + 3x_3 - 2x_4 = -9 \tag{10}$$

Add -3 times Equation (7) to Equation (9):

$$x_2 + 5x_3 + 2x_4 = -11 \tag{11}$$

36

We thus leave an equivalence system

$$x_1 + 2x_2 - x_3 + 4x_4 = 11$$
$$x_2 + 3x_3 - 2x_4 = -9$$
$$x_2 + 5x_3 + 2x_4 = -11$$

STEP 2.   Consider the system

$$x_2 + 3x_3 - 2x_4 = -9 \tag{10}$$
$$x_2 + 5x_3 + 2x_4 = -11 \tag{11}$$

Use Equation (10) to eliminate x_2 from Equation (11).
Add -1 times Equation (10) to Equation (11):

$$2x_3 + 4x_4 = -2 \tag{12}$$

Since there are no more variables that can be
eliminated, we move to the next step.

NOTE the equivalence system to solve is the
following:

$$x_1 + 2x_2 - x_3 + 4x_4 = 11 \tag{7}$$
$$x_2 + 3x_3 - 2x_4 = -9 \tag{10}$$
$$2x_3 + 4x_4 = -2 \tag{12}$$

STEP 3.   **Solve** for x_3 in terms of x_4 in Equation (12):

$$x_3 = -1 - 2x_4$$

Substitute $x_3 = -1 - 2x_4$ into Equation (10) and
solve for x_2:

$$x_2 + 3(-1-2x_4) - 2x_4 = -9$$
$$x_2 = -6 + 8x_4$$

Substitute $x_3 = -1 - 2x_4$ and $x_2 = -6 + 8x_4$ in Equation (7) and solve for x_1:

$$x_1 + 2(-6+8x_4) - (-1-2x_4) + 4x_4 = 11$$
$$x_1 = 22 - 22x_4$$

Therefore, the system has infinitely many solutions which are as follows:

x_4 = arbitrary
$x_1 = 22 - 22x_4$
$x_2 = -6 + 8x_4$
$x_3 = -1 - 2x_4$

The solutions can also be written as the solution set

$$\{(x_1,x_2,x_3,x_4) \mid x_4 \text{ is arbitrary}, x_1 = 22 - 22x_4,$$
$$x_2 = -6 + 8x_4, x_3 = -1 - 2x_4\}$$

EXAMPLE 3:   Solve

$$3x + 2y - 7z = 3 \qquad\qquad (14)$$
$$4x + 5y - 3z = 2 \qquad\qquad (15)$$
$$7x + 7y - 10z = 4 \qquad\qquad (16)$$

Solution:

STEP 1.   Since none of the variables have coefficient 1, it does not matter which variable we choose. The amount of computation involved will be the same. We choose Equation (14) to eliminate x from Equation (15) and Equation (16).

Multiple Equation (15) by -3:

$$-12x - 15y + 9z = -6 \qquad\qquad (17)$$

Multiply Equation (16) by (-3):

$$-21x - 21y + 30z = -12 \qquad (18)$$

Add 4 times Equation (14) to Equation (17):

$$-7y - 19z = 6 \qquad (19)$$

Add 7 times Equation (14) to Equation (18):

$$-7y - 19z = 9 \qquad (20)$$

The equivalent system to solve is:

$$3x + 2y - 7z = 3$$
$$-7y - 19z = 6$$
$$-7y - 19z = 9$$

STEP 2. Consider the system

$$-7y - 19z = 6 \qquad (19)$$
$$-7y - 19z = 9 \qquad (20)$$

Add -1 times Equation (19) and add to Equation (20). We obtain a false statement

$$0 = 3$$

which implies the system in inconsistant.

SELF-TEST:

1. A general system of linear equations may have

   a. _____
   b. _____
   c. _____

39

Ans:

a. a unique solution
b. infinitely many solutions
c. no solutions (inconsistant)

2. Solve the system by the method of elimination:

$$2x + y - 3z = -1 \quad \text{(a)}$$
$$3x + 5y - 2z = 0 \quad \text{(b)}$$
$$4x - 3y - 5z = 5 \quad \text{(c)}$$

STEP 1: We choose Equation ___ to eliminate the variable y from the remaining equations since the coefficient of y in that equation is 1.

Ans: (a)

To eliminate 5y from Equation (b), add ___ times Equation (a) to Equation (b):

Ans: -5

$$-7x + 13z = 5 \quad \text{(d)}$$

To eliminate -3y from Equation (c), add 3 times Equation (a) to Equation (b):

_____ (e)

Ans: $10x - 14z = 2$ (e)

List the equivalence system to be solved:

40

_____

_____

_____

$$2x + y - 3z = -1 \quad \text{(a)}$$
$$-7x + 13z = 5 \quad \text{(d)}$$
$$10x - 14z = 2 \quad \text{(e)}$$

STEP 2.   Consider the system

$$-7x + 13z = 5 \quad \text{(d)}$$
$$10x - 14z = 2 \quad \text{(e)}$$

To eliminate 10x from Equation (e), first multiply Equation (e) by 7 and obtain

_____ (f)

Ans: $70x - 98z = 14$ (f)

Add _____ times Equation (d) to Equation (f) and obtain

$$32z = 64 \quad \text{(f)}$$

Ans: 10

The equivalence system to solve is

$$2x + y - 3z = -1 \quad \text{(a)}$$
$$-7x + 13z = 5 \quad \text{(d)}$$
$$32z = 64 \quad \text{(f)}$$

STEP 3.   Solve for z in Equation (f):

$$z = \text{\_\_\_\_}$$

41

$$\underline{\text{Ans:}} \quad 2$$

Substitute z into Equation (d) to obtain the value for x:

$$x = \underline{\quad\quad}$$

$$\underline{\text{Ans:}} \quad 3$$

Substitute values for x, z into Equation (a) and obtain the value for y:

$$y = \underline{\quad\quad}$$

$$\underline{\text{Ans:}} \quad -1$$

The solution of the system is

$$(x,y,z) = (\underline{\quad},\underline{\quad},\underline{\quad})$$

$$\underline{\text{Ans:}} \quad (x,y,z) = (3,-1,2)$$

3. Solve the system by the method of elimination:

$$2x_1 + 5x_2 + x_3 = 5$$
$$3x_1 - 2x_2 + 5x_3 = -3$$
$$4x_1 + 4x_2 - x_3 = 19$$

$$\underline{\text{Ans:}} \quad (x,y,z) = (4,0,-3)$$

In Example 2, we have a system with solutions in which one of the variables can be assigned arbitrary values and the other variables are dependent on it. We will see from the next example that more than one variable may take arbitrary values.

<u>EXAMPLE 4:</u> Solve the system

$$-2x - 2y - 2z - w = -2 \qquad (21)$$
$$2x + 3y + 4z - 2w = 6 \qquad (22)$$
$$4x + 6y + 8z - 4w = 12 \qquad (23)$$

Solution: Add Equation (21) to Equation (22):

$$y + 2z - 3w = 4 \qquad (24)$$

Add 2 times Equation (21) to Equation (23)

$$2y + 4z - 6w = 8 \qquad (25)$$

Add -2 times Equation (24) to Equation (25)

$$0 = 0$$

We thus have the equivalent system

$$-2x - 2y - 2z - w = -2 \qquad (21)$$
$$y + 2z - 3w = 4 \qquad (24)$$

Solve for y in terms of z and w in Equation (24)

$$y = 4 - 2z + 3w$$

Substitute y = 4 - 2z + 3w into Equation (21):

$$-2x - 2(4-2z+3w) - 2z - w = -2$$
$$-2x - 8 + 4z - 6w - 2z - w = -2$$
$$-2x = -2z + 7w + 6$$
$$x = \tfrac{1}{2}(2z-7w-6)$$

Therefore, the solutions of the system are

$$w = \text{arbitrary}$$
$$z = \text{arbitrary}$$
$$y = 4 - 2z + 3w$$
$$x = \tfrac{1}{2}(2z-7w-6)$$

43

Given any value for w and any value for z, a solution of
the system can be obtained. For example, w = 0, z = 1,
y = 2, x = -2 is a solution, and (x,y,z,w) = (5,-4,1,-2) is
also a solution.   The solution set can be written as

$$A = \{(x,y,z,w) \mid w,z \text{ are arbitrary,}$$
$$y = 4 - 2z + 3w, \ x = \tfrac{1}{2}(2z-7w-6)$$

If you solve for z in terms of y and w in Equation (24)
we obtain

$$z = \tfrac{1}{2}(4-y+3w)$$

Substitute z in Equation (21) and solve for x:

$$x = -\tfrac{1}{2}(2+y+4w)$$

Therefore, the solutions are

$$y = \text{arbitrary}$$
$$w = \text{arbitrary}$$
$$z = \tfrac{1}{2}(4-y+3w)$$
$$x = -\tfrac{1}{2}(2+y+4w)$$

The variables that take arbitrary values are y and w instea
of the variables w and z.   However the solution set

$$B = \{(x,y,z,w) \mid y,w \text{ are arbitrary, } z = \tfrac{1}{2}(4-y+3w),$$
$$x = -\tfrac{1}{2}(2+y+4w) \}$$

is the same as the solution set A.

SELF-TEST: In the above example, find a value for y and a
value for w so that (5,y,1,w) is an element of B, thus a
solution of the system.

<u>Ans:</u> y = -4, w = -2

If we let y = 6, w = 2 in the solution set B, then

44

$(x,y,z,w) = (-8,6,2,2)$ is a solution.

Find values for w and z so that $(-8,6,z,w)$ is an element of A.

<u>Ans:</u>   z = 2, w = 2

If a system has more variables than equations and if it has a solution, then it must have infinitely many solutions.

<u>EXAMPLE 5</u> : Solve the system

$$x + y + z = 3$$
$$2x - y + 3z = 4$$

<u>Solution:</u> By inspection, the system has a solution $x = y = z = 1$. Since the system has 3 variables which is a number greater than 2 (the number of equations) we know the system must have infinitely many solutions and the solutions are

$$y = \text{arbitrary}$$
$$x = 5 - 4y$$
$$z = -2 + 3y$$

<u>NOTE:</u> 1. If the number of equations is more than (or the same as) the number of variables in a system, then the system may have one, infinitely many or no solutions.

2. A system with a number of equations less than the number of variables may have no solutions. For example, the system

$$2x + y - z = 3$$
$$4x + 2y - 2z = 5$$

has no solutions.

45

1. If a system of linear equations with more variables than the number of equations has a solution, then it must have _____.

    Ans: Infinitely many solutions

2. The system

    $$x + y + 2z = 0$$
    $$3x - 2y - z = 0$$

    has a trivial solution $(x,y,z) = (\_\_,\_\_,\_\_)$.

    Ans: $(x,y,z) = (0,0,0)$

    Thus it must have _____.

    Ans: Infinitely many solutions.

    This is an example of a homogeneous system.

Applications:

EXAMPLE 6: The sum of three integers is 14. Find the integers if the sum of the first two is the third and twice the third is 4 more than the second.

Solution: Let x, y, z denote the first, second and third integers respectively. Their sum is 14. Thus,

$$x + y + z = 14$$

Also we have

$$x + y = z$$
$$2z = 4 + y$$

To solve the system

$$x + y + z = 14 \qquad\qquad (18)$$
$$x + y - z = 0 \qquad\qquad (19)$$
$$-y + 2z = 4 \qquad\qquad (20)$$

We use the method of elimination. Use Equation (18) to eliminate y from Equation (19). Add -1 times Equation (18) to Equation (19):

$$-2z = -14$$
$$z = 7 \qquad\qquad (21)$$

Substitute z in Equation (21) into Equation (20). We have

$$y = 10 \qquad\qquad (22)$$

Substitute z in Equation (21) and y in Equation (22) into Equation (18). We have

$$x = -3$$

Therefore, the integers are -3, 10 and 7.

EXAMPLE 7: Three pounds of coffee, 4 pounds of sugar and 6 pounds of flour cost $25.60. Two pounds of sugar and 3 pounds of flour cost 55¢ more than 2 pounds of coffee. Ten pounds of coffee cost $4 more than 10 pounds of sugar and 10 pounds of flour combined. How must does each pound of coffee, sugar and flour cost?

Solution: Let

x = the cost of coffee per pound

y = the cost of sugar per pound

z = the cost of flour per pound

It follows that

the cost of 3 pounds of coffee = 3x

the cost of 4 pounds of sugar = 4y

the cost of 6 pounds of flour = 6z

Since the total cost is $25.60, we thus have the equation

$$3x + 4y + 6z = 25.60$$

The cost of

2 pounds of sugar = 2y

3 pounds of flour = 3z

2 pounds of coffee = 2x

The condition that 2 pounds of sugar plus 3 pounds of flour cost 55¢ more than 2 pounds of coffee give us

$$2y + 3z = 2x + 0.55$$

Finally, the cost of

10 pounds of coffee = 10x

10 pounds of sugar = 10 y

10 pounds of flour = 10z

and the fact that 10 pounds of coffee is $4 more than 10 pounds of sugar and 10 pounds of flour together gives us

$$10 x = 10y + 10z + 4$$

Thus, we have to solve the system

$$3x + 4y + 6z = 25.60$$
$$2y + 3z = 2x + 0.55$$
$$10x = 10y + 10z + 4$$

This system may be rewritten as follows:

$$3x + 4y + 6z = 25.60 \tag{23}$$
$$-2x + 2y + 3z = 0.55 \tag{24}$$
$$10x - 10y - 10z = 4 \tag{25}$$

To solve the system, we first use Equation (24) to eliminate y in Equation (23) and Equation (25) by adding -2 times Equation (24) to Equation (23):

$$7x = 24.50$$
$$x = 3.50$$

and adding 5 times Equation (24) to Equation (25):

$$5z = 6.75$$
$$z = 1.35$$

Substitute $x = 3.50$ and $z = 1.35$ in Equation (24) to obtain

$$y = 1.75$$

Therefore, coffee costs $3.50 a pound, sugar $1.75 a pound and flour $1.35 a pound.

EXAMPLE 8: A 50 pound bag of fertilizer contains nitrogen, phosphate and potash. It is known that for every 3 pounds of nitrogen there is 1 pound of phosphate and for every 4 pounds of nitrogen there are 3 pounds of potash. How much nitrogen is contained in the bag? Phosphate? Potash?

Solution: Let the bag contain

x pounds of nitrogen

y pounds of phosphate

z pounds of potash

Since it is a 50 pound bag, we have

$$x + y + z = 50$$

Knowing that for every 3 pounds of nitrogen, there **is** 1 pound of phosphate we have the relation

$$x = 3y$$

The information that for every 4 pounds of nitrogen there are 3 pounds of potash tells us that 3 times the total weight of nitrogen is the same as 4 times the total weight of potash, thus

$$3x = 4z$$

The system to solve is

$$x + y + z = 50 \tag{26}$$
$$x - 3y = 0 \tag{27}$$
$$3x - 4z = 0 \tag{28}$$

Solve for y in Equation (27): $y = \dfrac{x}{3}$

Solve for z in Equation (28): $z = \dfrac{3x}{4}$

Substitute in Equation (26):

$$x + \frac{x}{3} + \frac{3x}{4} = 50$$
$$12x + 4x + 9x = 600$$
$$x = 24$$

hence

$$y = \frac{24}{3} = 8$$

$$z = \frac{3}{4} \cdot 24 = 18$$

Therefore, the 50 pound bag of fertilizer contains 24 pounds of nitrogen, 8 pounds of phosphate and 18 pounds of potash.

EXAMPLE 9: A car dealer ordered 4 different sizes of cars to fill his parking lot. The lot size is 10,000 square feet, of which 3,155 square feet must be used for driving and spaces between cars. The footages and prices of the cars ordered are given below:

	FOOTAGE	PRICE
subcompact	54 ft^2	$4,500
compact	62.5 ft^2	$5,000
medium size	71.5 ft^2	$6,000
large size	84 ft^2	$7,500

Knowing that small cars are in great demand, the dealer ordered 10 less subcompact cars than the total of the other three sizes. The cost for all subcompact cars is 3 times that for all the large size cars. If the total cost for all the cars ordered is $570,000, how many of each size of cars did the dealer order?

Solution: Let

    x = the number of subcompact cars ordered
    y = the number of compact cars ordered
    z = the number of medium size cars ordered
    w = the number of large size cars ordered

The spaces occupied by each size of cars are as follows:

    subcompact   54x ft^2
    compact   62.5y ft^2

51

medium size   71.5z ft²
large size   84w ft²

Since the total available parking space is 10,000 - 3155 = 6845 f which is to be filled,

$$54x + 62.5y + 71.5z + 84w = 6845$$

Now, we know the dealer ordered 10 less subcompact cars than the combination of the other three, this gives us the relation

$$y + z + w = 10 + x$$

The costs for the different sizes of cars are

subcompact   4,500x
compact   5,000y
medium size   6,000z
large size   7,500w

and the total is $570,000. Hence,

$$4500x + 5000y + 6000z + 7500w = 570000$$

Finally we know

$$4500x = 3(7500w)$$

The system to solve is:

$$54x + 62.5y + 71.5z + 84w = 6845$$
$$-x + y + z + w = 10$$
$$4500x + 5000y + 6000z + 7500w = 570000$$
$$4500x - 22500w = 0$$

Solve the system and we obtain the unique solution

x = 50, number or subcompact cars ordered

y = 30, number of compact cars ordered

z = 20, number of medium size cars ordered

w = 10, number of large size cars ordered

EXAMPLE 10: A 15 pound food product contains three different ingredients A, B, C. Their contents of nutrient P and nutrient Q per pound of ingredient are given as follows:

	Nutrient P	Nutrient Q
Ingredient A	5 ounces	4 ounces
Ingredient B	3 ounces	3 ounces
Ingredient C	1 ounce	2 ounces

It is required that the food product must contain 71 ounces of Nutrient P and 58 ounces of Nutrient Q. How much of each ingredient A, B, C should be contained in the food product?

Solution: Let

x be the amount in pounds of ingredient A

y be the amount in pounds of ingredient B

z be the amount in pounds of ingredient C

The total contents in ounces of Nutrient P and Nutrient Q contained ingredients A, B, C are thus as follows:

	Nutrient P	Nutrient Q
Ingredient A	5x	4x
Ingredient B	3y	3y
Ingredient C	z	2z

We then have the following system to solve:

$$x + y + z = 15$$
$$5x + 3y + z = 71$$
$$4x + 3y + 2z = 58$$

The system is equivalent to

$$x + y + z = 15$$
$$-y - 2z = -2$$

and the solutions are

$$z = \text{arbitrary}$$
$$y = 2 - 2z$$
$$x = 13 + z$$

The system has infinitely many solutions. However, in order
to have the answer meaningful, we must have x, y, z all positive.
One possible answer is $z = 0.5$, $y = 1$, $x = 13.5$, i.e., the
food product contains 13.5 pounds of Ingredient A, 1 pound of
Ingredient B and half a pound of Ingredient C. On the other
hand, the solution $z = 2$, $x = 15$, $y = -2$ for the system
cannot be a true statement for the problem.

SELF-TEST: Three dozen eggs, 5 pounds of beef and 4 chickens
cost altogether $27.82. Each chicken costs 80¢ less than a
pound of beef. John received $4 change from $15 for buying
two pounds of beef and 2 chickens. How much does each dozen
eggs, one pound of beef and one chicken cost?

Ans:

1 dozen eggs = 89¢
1 pound of beef = $3.15
1 chicken = $2.35

A linear equation in three variables is represented by a
plane in space. Therefore, to find points which lie on two
planes in space one can solve a system of two linear
equations in three variables. Note such a system either has
no solution, in which case the two planes do not intersect; or
has infintely many solutions, in which case they are either
the same plane or intersect at a line. Consider the following

example.

EXAMPLE 11: Determine if the two planes P_1 and P_2 intersect at a line, coincide or do not intersect.

$$P_1: \quad 2x + y - z = 3$$
$$P_2: \quad 3x - 2y + 3z = 1$$

Solution: Solve the system to obtain the solution

$$x = r, \text{ arbitrary}$$
$$y = 10 - 9r$$
$$z = 7 - 7r$$

which are the parametric equations for the line of the inter-section of planes P_1 and P_2.

SELF-TEST: Determine if P_3 and P_4 intersect at a line, coincide or do not intersect.

$$P_3: \quad 2x - 3y + z = -1$$
$$P_4: \quad -4x + 6y - 2z = 3$$

Ans: P_3 and P_4 do not intersect.

Homogeneous System of linear equations:

If the constant term of each equation in a system is zero, the system is called a homogeneous system. Thus, a homo-geneous system can be expressed as

$$a_{11}x_1 + a_{12}x_2 + \ldots + a_{1n}x_n = 0$$
$$a_{21}x_1 + a_{22}x_2 + \ldots + a_{2n}x_n = 0$$
$$\vdots \qquad \vdots \qquad \qquad \vdots \qquad \vdots$$
$$a_{m1}x_1 + a_{m2}x_2 + \ldots + a_{mn}x_n = 0$$

55

For example,

$$3x + 2y - z = 0$$
$$2x - y + 5z = 0$$
$$-x - y = 0$$

is a homogeneous system of three equations in three unknowns.

A homogeneous system of linear equations always has a trivial solution, namely the solution in which the values of all the variables are zero, i.e., $x_1 = x_2 = \ldots = x_n = 0$. Therefore, given a homogeneous system, it is more interesting to find out if it has a non-trivial solution; i.e., a solution in which at least one of the variables is non-zero. In fact, if a homogeneous system has one non-trivial solution, then it must have infinitely many solutions since a system can not have only two solutions.

EXAMPLE 12: The system

$$3x + 2y = 0$$
$$2x - 5y = 0$$

has only the trivial solution $(x,y) = (0,0)$

EXAMPLE 13: The system

$$x + y - z = 0$$
$$2x - 2y + 3z = 0$$
$$3x - y + 2z = 0$$

has infinitely many solutions:

$$z = r, \text{ arbitrary}$$
$$y = \frac{5}{4}r$$
$$x = \frac{-1}{4}r$$

56

It is important to note that if the number of variables is more than the number of equations, then the homogeneous system will have infinitely many solutions. This is because a homogeneous system always has a non-trivial solution and, thus, has infintely many solutions for having more variables than equations.

SELF-TEST:

1. If the constant term of each equation is zero, the system is called _____.

   > Ans: homogeneous

2. A homogeneous system always has a _____ solution.

   > Ans: trivial or
   > zero

3. If a homogeneous system has a non-trivial solution (non-zero) then it will have _____.

   > Ans: infinitely many
   > solutions

4. If a homogeneous system has more variables than equations, then it will have _____.

   > Ans: infinitely many
   > solutions

5. Determine if the following system has non-trivial solutions.

$$3x - 5y + z = 0$$
$$5x - 7y + 5z = 0$$
$$-x + 3y + 3z = 0$$

   > Ans: yes

EXERCISE 1.3

Solve the systems 1-26.

1. $\begin{aligned} x + 2y - z &= 4 \\ 2x - y + 3z &= -3 \\ 3x + y + z &= 4 \end{aligned}$

2. $\begin{aligned} x + 3y - z &= 1 \\ 2x - 2y + z &= 2 \\ -x + y + 2z &= 3 \end{aligned}$

3. $\begin{aligned} t + 3u + v &= 0 \\ 2t - 4u - v &= 3 \\ 3t + u + 2v &= 1 \\ u + v &= -1 \end{aligned}$

4. $\begin{aligned} r - s + 2t &= -1 \\ 2r + 2s - t &= 0 \\ r + s &= 5 \\ 2r + 2t &= 4 \end{aligned}$

5. $\begin{aligned} x + 2y - 3z &= -1 \\ 2x - 3y + 4z &= 2 \end{aligned}$

6. $\begin{aligned} 2x + y - 2z &= 2 \\ 6x + y - 5z &= 1 \end{aligned}$

7. $\begin{aligned} 2x + 3y - 4w &= -1 \\ 4x + 6y - 8w &= 2 \end{aligned}$

8. $\begin{aligned} r + s - u + 3w &= -1 \\ 3r - s + 2u + 5w &= 3 \\ 4r + u + 8w &= 5 \end{aligned}$

9. $\begin{aligned} x_1 - x_2 - 3x_3 - x_4 &= 0 \\ x_1 + 2x_2 - 4x_3 - 5x_4 &= -3 \\ 2x_1 - 2x_2 + 3x_3 - 4x_4 &= 4 \\ 3x_1 - x_2 - x_3 + 4x_4 &= 11 \end{aligned}$

10. $\begin{aligned} 3u + 2v + 4w - 5x &= -1 \\ u - v + 2w - x &= 2 \\ 2u + 2v - 3w + 4x &= 3 \\ -u + 2v - 3w + x &= -3 \end{aligned}$

11. $\begin{aligned} x + 2y + z &= 0 \\ -2x - y + 3z &= 0 \\ 3x - y + 4z &= 0 \end{aligned}$

12. $\begin{aligned} 2u - v + 3w &= 0 \\ u + 2v - 2w &= 0 \\ 3u - 3v + 2w &= 0 \end{aligned}$

13.
$$x + y + z = 0$$
$$2x - 2y + 2z = 0$$

14.
$$u + v = 0$$
$$2u - 2v = 0$$
$$3u - v = 0$$

15.
$$2x + 3y + 4z - w = +1$$
$$x - w + z = 2$$

16.
$$u + 2v - w = 3$$
$$4u + 5v = -2$$

17.
$$x_1 + 2x_2 + 3x_3 = 1$$
$$2x_1 + 3x_2 + 4x_3 = 2$$
$$3x_1 + 4x_2 + 5x_3 = 3$$

18.
$$x_1 + x_2 + x_3 = 1$$
$$2x_1 + 2x_2 + 2x_3 = -1$$
$$x_1 - x_2 + x_3 = 3$$

19.
$$x_1 + x_2 + x_3 = 1$$
$$2x_1 + 2x_2 + 2x_3 = 2$$
$$3x_1 + 3x_2 + 3x_3 = 3$$
$$4x_1 + 4x_2 + 4x_3 = 4$$

20.
$$u + v - w = 2$$
$$3u + 3v - 3w = 6$$
$$9u + 9v - 9w = 18$$

21.
$$x + y - z + w = -1$$
$$x + 2w = 0$$

22.
$$x_1 + 2x_2 - 3x_3 = 4$$
$$x_2 - x_4 = 5$$

23.
$$2x_1 - x_2 - x_3 + 4x_4 + x_5 = 4$$
$$3x_1 + 2x_2 + x_3 + 5x_4 - x_5 = 2$$
$$x_1 + x_2 - x_3 + 3x_4 + x_5 = 5$$
$$-x_1 + 2x_2 - 2x_3 + 6x_4 - 2x_5 = -1$$
$$-2x_1 - 3x_2 + x_3 - 5x_4 + x_5 = -4$$

24.
$$u + 3v - 4w + x - y = 6$$
$$2u - v + 2w + 3x - 2y = -5$$
$$4u - 3v - 2w + 5x + y = 5$$
$$-2u + v + w + 2x + 2y = 2$$
$$3u - 2v - 3w + 5x + y = 6$$

25.
$$0.2x - 0.3y + 1.5z = 1.3$$
$$0.1x + 0.15y + z = 3.3$$
$$2x + 0.2y + 0.1z = 2.7$$

26. $0.5u + 3.4v - 4.1w = -4.6$
    $u + 4v - w = -2$
    $2u + 3.1v + 5.2w = 3.2$

27. Find a condition on a, b, c, d such that the system

    $ax + by = 0$
    $cx + dy = 0$

    has only the trivial solution. (HINT: See Problem 13 in section 1.2)

28. Explain why the following system has infintely many solutions:

    $3x + 5y = 0$
    $-2x + 2z + w = 0$
    $x - 3z = 0$

29. The sum of three positive numbers is 90 and the sum of two of them is the third. Find the numbers if the smallest one is half of the largest.

30. The sum of three numbers is 10.3. Twice the first is 0.1 less than the second and three times the third number is 6.2 greater than the sum of the other two. What are the three numbers?

31. Diana bought 4 lbs. of coffee, 5 lbs. of sugar and 10 lbs. of flour and received $2.25 change from $30. Nine lbs. of flour is 50¢ more than 5 lbs. of sugar. Ten pounds of flour and 4 lbs. of sugar cost $2 more than 3 lbs. of coffee. Find the price for each pound of coffee, sugar and flour.

32. John spent $20 buying Christmas gifts for his parents

60

and sister Mary. Twice the money he spent on the gift
for Mary is $9 less than that for his parents. Three times
the money spent on his mother's gift is $2 more than the
combination of twice the money on his father's gift and
four times Mary's. How much money did John spend on
buying each gift?

33. Wendy received $2.10 in change which is in dimes, nickles
and quarters. Suppose eggs cost 85¢ a dozen and milk costs
$1.95 a gallon. The quarters and dimes she received can
buy a gallon milk and twice the nickles and quarters
received can buy a dozen eggs and a gallon of milk. How
many nickles, dimes and quarters did Wendy receive?

34. The perimeter of an isosciles trapezoid is 25 feet. The sum
of the two bases is two feet longer than the sum of the two
sides and four times a side is twice the longest base.
Find the length of the two bases and the sides.

35. A real estate company rented an office which is 50 feet
long and 40 feet wide. Forty desks are moved into the
office. The sizes of desks are given below:

	Length	Width
small size	3'	3'
medium size	3'	4'
large size	3'	5'

It is known that 1565 square feet of office space is
used for purposes other than placing desks. If there are
the same number of small desks as the combination of
medium and large desks, how many desks of each are moved
in the office?

36. David invested $5,000 buying three stocks A, B, C. The

cost per share and annual return of each stock are given below:

	Cost per share	Annual return
Stock A	$15	7%
Stock B	$21	8%
Stock C	$10	6%

The money he spent on Stock B is $500 less than the total in buying Stock A and ten times that of Stock C. At the end of one year David received $378 in return. How many shares of each stock did he invest in?

37. A manufacturing company produces four products A, B, C, D which require three machines to process. The time needed to produce one unit of the products is as follows:

	A	B	C	D
Machine I	1	3	2	4
Machine II	2	4	7	2
Machine III	1	5	1	3

The available machine time per week is:

Machine I	110 hours
Machine II	120 hours
Machine III	115 hours

Find the number of units of each product produced each week if all machines are operated to their full capacity.

38. Let A be a four digit positive integer. If the digit in the thousandth place is switched with the digit in the unit place, the number will be increased by 999 and if the

62.

hundredth place is interchanged with the 10th place, the number will be 360 less. If the four digits are reversed, the number will be 639 'larger. Find the integer if the sum of four digits is 13.

39. Ten years ago, Joe was 3 years older than his brother Mike and 7 years older than his sister Jane. Five years from now, Joe's mother will be twice as old as her daughter Jane. Assume this is year 1982, then the combined ages of Joe, his brother Mike, his sister Jane, and their mother will be 278 in the year 2000, assuming they will all be alive. How old are they?

40. The balance of an account is $251 more than it should be. The error is caused from three items. The amount of item I is a two digit integer and is $23 more than the correct amount due to digits reversing. Item II has an amount with three digits and exceeds its actual amount by $148 also due to reversing the digits. Item III has an amount of 2 digits with unit digit 0 and is mistakenly increased by $30. What is the amount of each item?

REVIEW OF CHAPTER ONE:

1. Write down the general form of a system of two linear equations in two unknowns.

2. A solution of a system of two equations in two unknowns is _____.

3. To solve a system is to find _____.

4. The methods given in this chapter to solve a system of two linear equations in two unknowns are called the _____.

5. Operations used in the method of elimination to solve a system of two equations and two variables are _____ and _____.

6. A system of linear equations may have

   a. _____ solutions.
   b. _____
   c. _____

7. If a system has more than one solutions, it must have _____ solutions.

8. To find the point of intersection of two linear equations is the same as _____.

9. If the system of two linear equations represented by two lines have only one solution, then the lines are _____. And if the system has infinitely many solutions, the lines are _____ _____. And if the system has no solution, the lines are _____.

10. Two systems are said to be equivalent if _____ _____.

11. A system that has no solution is said to be _____.

12. A general system of m linear equations in n variables can be given as _____.

13. A general system of linear equations may have
    (a) _____ (b) _____ (c) _____.

14. The most commonly used method in solving a system of linear equations is _____.

15. The operations allowed to change a system to an equivalent one are

        a. _____

        b. _____

        c. _____

16. If a system has more variables than equations, then either it has no solution or _____.

17. A homogeneous system is one in which the constant terms of each equation are _____.

18. A homogeneous system of linear equations always has a _____ solution.

19. If the number of equations is less than the number of variables in a homogeneous system then the system will have _____.

20. Two distinct planes in space either do not intersect or intersect at a _____.

21. To find points of intersection of two planes in space is the same as solving a system of _____ equations in _____ variables.

SAMPLE TEST: CHAPTER ONE

1. Solve the system

   a. $5x - y = 3$
      $2x + 3y = 1$

   b. $3u + 2v = 5$
      $7u - 3v = -2$

2. Determine if the lines L_1 and L_2 given below have one point of intersection, do not intersect or are the same line.

   a. L_1: $3x - 4y = -2$
      L_2: $5x + 6y = 7$

   b. L_1: $-3x + 2y = 3$
      L_2: $6x - 4y = 5$

3. Solve each of the systems:

   a. $2x_1 - x_2 + 3x_3 = -14$
      $3x_1 + 2x_2 + x_3 = 7$
      $4x_1 + 3x_2 - 2x_3 = 27$

   b. $x + 2y - 3z - w = 3$
      $3x - y + 2z + 2w = -5$
      $4x + y - z + w = -2$

   c. $2x + y - 3z = -1$
      $3x - 5y + z = 2$
      $-x - 7y + 7z = 4$

   d. $x_1 + x_2 + x_3 + x_4 = 0$
      $2x_1 + 3x_2 - x_3 + 2x_4 = 0$
      $5x_1 - 2x_2 + 2x_3 - x_4 = 0$

4. Determine if the planes P_1 and P_2 intersect in a line, coincide (same plane) or do not meet.

   P_1: $2x + 3y - z = 4$
   P_2: $3x - 2y + 5z = 1$

66

5. Find the number of rabbits and chickens if their total number is 12 and the total number of legs is 34.

6. Tom invested $2,000 in buying two stocks. The annual return from one stock is 11% and from the other it is 10%. If he received $212 in return from both stocks at the end of one year, how much did he invest in each stock?

7. Mrs. Smith bought 8 pounds of ham, 6 pounds of beef and 10 pounds of chicken and paid $59. If she bought only 1/4 of the ham, 1/3 of the beef and half of the chicken, she would have had to pay $18.75. Knowing that beef is only 25¢ per pound less than ham, how much does each pound of ham, beef and chicken cost?

8. A shoe factory makes four different quality shoes. The cost of each grade of shoe is given below:

	Cost per pair
Grade A	$35
Grade B	$27
Grade C	$15
Grade D	$8.5

A retail shoe store sent in a check for $1135 and indicated they would like to spend all the money buying different grades of shoes such that the number Grade A Shoes equals the number of Grade B shoes. The number of Grade D shoes ordered should be three times that of Grade A and the number of Grade C should be twice that that of A. Can you figure out how many pairs of shoes from each grade the store would like to order?

COMPUTER APPLICATIONS: CHAPTER ONE

1. Write a BASIC program to do the following:

    Read in 100 pairs of lines and determine if each pair of lines intersect at one point, are parallel or the same line. If they intersect at a point, print out the point.

2. Write a BASIC program to do the following:

    Read in 100 pairs of planes and determine if each pair of planes intersect at a line are the same plane or do not meet. If the planes intersect in a line, give the line of intersection.

3. a. Draw a flow chart for the Gaussian elimination method to solve a system of linear equations.

    b. Write a BASIC program to solve a system of linear equations by the method of elimination.

4. Write a BASIC program to do the following:

    a. Read in 100 3-digit positive integers.

    b. Count the number of integers which are greater than the number obtained by reversing the original.

    c. Print out the original list, the list with digits reversed for each integer, also the number obtained from b.

CHAPTER TWO

MATRICES

2.1 DEFINITION OF A MATRIX AND SPECIAL MATRICES

OBJECTIVES:

1. *Define a matrix and introduce some notations used for matrices.*
2. *Introduce some special matrices such as square, symmetric, triangular, and diagonal matrices.*

DEFINITION 2.1: A matrix is a rectangular array of numbers.

Examples of matrices are:

$$A = \begin{bmatrix} 1 & -5 \\ 3 & 0 \end{bmatrix}, \quad B = \begin{bmatrix} 2 & \frac{1}{2} & -1 \\ 3 & 7 & \frac{4}{3} \end{bmatrix},$$

$$C = \begin{bmatrix} 4.1 & 3.2 & 2 \\ 5.4 & -1 & 0 \\ 9.5 & 4 & 5.2 \end{bmatrix}$$

The numbers of the matrix are called the entries. In the above examples, the matrix A has two rows and two columns. The number 1 is called the (1,1) - entry of A since 1 appears in the first row and first column, the (1,2)-entry is -5 which

appears in the first row and second column. Similarly, the
(2,1) - entry and (2,2) - entry of the matrix A are 3 and
0 respectively. The matrix A is said to be a 2 × 2 (two by
two) matrix for having two rows and two columns which determine
the order of the matrix A.

Accordingly, the matrix B is a 2 × 3 matrix, hence of
order 2 × 3, and C is a matrix of order 3 × 3. In general,
a matrix A with m rows and n columns or an m × n matrix is
written as:

$$
A = \begin{bmatrix} a_{11} & a_{12} & --- & a_{1n} \\ a_{21} & a_{22} & --- & a_{2n} \\ a_{m1} & a_{m2} & --- & a_{mn} \end{bmatrix}
$$

or abbreviated as

$$
A_{m \times n} = \begin{bmatrix} a_{ij} \end{bmatrix}_{m \times n}
$$

which means A is a matrix with m rows and n columns whose
(i,j) - entry is a_{ij}.

Matrices may be used to store and represent information.

EXAMPLE 1. A mileage chart between major cities in the U.S.
can be represented by the following matrix:

	Atlanta	Boston	Chicago	Dallas	Los Angeles	New York
Atlanta	0	1074	703	813	2209	869
Boston	1074	0	993	1817	3057	214
Chicago	703	993	0	938	2104	845
Dallas	813	1817	938	0	1406	1604
Los Angeles	2209	3057	2104	1406	0	2837
New York	869	214	845	1604	2837	0

The (i,j)-entry of the matrix represents the distance

70

between city i and city j. Thus, the (4,2)-entry is the distance between Dallas and Boston which is the same as the (2,4)-entry, the distance between Boston and Dallas.

EXAMPLE 2. The following 4×2 matrix represents the area and populations of four continents:

	Area in sq. miles	Population
Africa	11,700,000	248,653,000
North America	8,440,000	261,348,000
South America	6,900,000	137,000,000
Australia	2,948,000	10,158,000

EXAMPLE 3. Inventory matrix:

	Warehouse 1	Warehouse 2	Warehouse 3
Sofa	5	3	10
Chair	15	25	12

EXAMPLE 4. Selling prices of grocery items of different supermarkets can be expressed in matrix form.

	Supermarket I	Supermarket II	Supermarket III
eggs	0.89	0.98	0.75
sugar	3.50	3.25	3.75
flour	2.50	2.15	2.65

SELF TEST

a. Let

$$A = \begin{bmatrix} 1 & -1 \\ 2 & 3 \\ 0 & 4 \end{bmatrix}$$

The matrix A has ___ rows and ___ columns and is called

a _____ matrix. The (1,2)-entry of A is _____ and the (3,2)-entry of A is _____.

b. The notation $B = [b_{ij}]_{r \times s}$ means B is a matrix with ___ rows and ___ columns. The (m,n)-entry of B is _____.

SPECIAL MATRICES:

1. <u>Square matrix:</u> A matrix which has the same number of rows as columns is called a square matrix. Examples of square matrices are:

$$\begin{bmatrix} 1 & -1 \\ 2 & 0 \end{bmatrix} \qquad \begin{bmatrix} 1 & 0 & 0 \\ 0 & 1 & 0 \\ 0 & 0 & 1 \end{bmatrix} \qquad \begin{bmatrix} 0 & 0 & 1 & 0 \\ 0 & 0 & 0 & 0 \\ 0 & 1 & 0 & 0 \\ 0 & 0 & 1 & 1 \end{bmatrix}$$

If $A = [A_{ij}]_{n \times n}$ is a square matrix, the main diagonal of A consists of a_{11}, a_{22}, ---, a_{nn}. For example, the main diagonal of

$$A = \begin{bmatrix} 1 & 2 \\ 4 & 5 \end{bmatrix}$$

consists of 1,5.

2. <u>Symmetric matrix:</u> A matrix is called symmetric if the (i,j)-entry is equal to the (j,i)-entry for each i,j. A symmetric matrix must be square. Examples of symmetric matrices are:

$$\begin{bmatrix} 1 & -1 \\ -1 & 2 \end{bmatrix} \quad \begin{bmatrix} 1 & 0 & 3 \\ 0 & -1 & 5 \\ 3 & 5 & 2 \end{bmatrix} \quad \begin{bmatrix} 2 & 2 \\ 2 & 2 \end{bmatrix}$$

3. Triangular matrix: A square matrix is called upper (lower) triangular if all the entries below (above) the main diagonal are zero. Examples of upper triangular matrices are

$$\begin{bmatrix} 1 & 2 \\ 0 & 3 \end{bmatrix} \quad \begin{bmatrix} 1 & -1 & 2 \\ 0 & 3 & 4 \\ 0 & 0 & -1 \end{bmatrix}$$

Examples of lower triangular matrices are:

$$\begin{bmatrix} -1 & 0 \\ 1 & 3 \end{bmatrix} \quad \begin{bmatrix} 1 & 0 & 0 \\ 2 & -1 & 0 \\ -1 & 1 & 3 \end{bmatrix}$$

4. Diagonal matrix and scalar matrix: A square matrix is called diagonal if all entries above and below the main diagonal are zero. In other words, a diagonal matrix is a matrix which is both upper and lower triangular. Examples of diagonal matrices are:

$$\begin{bmatrix} 1 & 0 \\ 0 & 2 \end{bmatrix} \quad \begin{bmatrix} 1 & 0 & 0 \\ 0 & 0 & 0 \\ 0 & 0 & 3 \end{bmatrix}$$

If all entries on the main diagonal of a diagonal matrix are equal, then the matrix is called scalar. Examples of scalar matrices are:

$$\begin{bmatrix} 2 & 0 \\ 0 & 2 \end{bmatrix} \quad \begin{bmatrix} 3 & 0 & 0 \\ 0 & 3 & 0 \\ 0 & 0 & 3 \end{bmatrix} \quad \begin{bmatrix} 1 & 0 & 0 \\ 0 & 1 & 0 \\ 0 & 0 & 1 \end{bmatrix}$$

The scalan matrix

$$\begin{bmatrix} 1 & 0 & 0 \\ 0 & 1 & 0 \\ 0 & 0 & 1 \end{bmatrix}$$

has all 1's on the main diagonal and is called the
(3 x 3) identity matrix. The identity matrix has properties
similar to those of the real number 1.

5. Column matrix and row matrix: A matrix is called a
column matrix if it consists of one column and is called
a row matrix if it consists of one row. Examples of
column matrices and row matrices are:

$$\begin{bmatrix} 1 \\ 2 \end{bmatrix} \quad \begin{bmatrix} 1 \\ -1 \\ 0 \end{bmatrix} \quad \begin{bmatrix} 2 & 1 & 0 \end{bmatrix} \quad \begin{bmatrix} 1 & -1 & 1 \end{bmatrix}$$

Column matrices and row matrices are somtimes called
column vectors or row vectors.

SELF-TEST

Give an example of each of the following:

- (a) a square matrix
- (b) a symmetric matrix
- (c) a triangular matrix
- (d) a diagonal matrix
- (e) a scalar matrix
- (f) a column matrix
- (g) a row matrix

EXERCISES 2.1

1. Let

$$A = \begin{bmatrix} 2 & -5 & 0 & 4 \\ -1 & 3 & 1 & 2 \\ 3 & -3 & 2 & 6 \end{bmatrix}$$

$$B = \begin{bmatrix} 4 & 2 & -1 \\ 3 & 0 & 9 \end{bmatrix}$$

   (a) Find the order of Matrix A.
   (b) Find the order of Matrix B.
   (c) Find the (2,3)-entry and the (3,3)-entry of Matrix A.
   (d) Find the (1,2)-entry and the (2,3)-entry of Matrix B.

2. Let

$$A = \begin{bmatrix} 1 & x \\ 5 & 4 \end{bmatrix}, \qquad B = \begin{bmatrix} y & -1 \\ 3 & 4 \end{bmatrix}$$

   (a) If the (1,2)-entry of A is the sum of the (2,1) and (2,2)-entries of B, find the value for x.
   (b) The (1,1)-entry of A is the sum of the (1,1) and (1,2)-entries of B. Find the value of y.

3. Let $C = [C_{ij}]_{mxn}$

   (a) What is the order of C?
   (b) What is the (4,3)-entry of C?
   (c) What is the $(\imath, \jmath)$-entry of C?

4. Let

$$A = \begin{bmatrix} 1 & 4 \\ -2 & 3 \end{bmatrix} \qquad B = \begin{bmatrix} 2 & 2 \\ -1 & 0 \end{bmatrix} \quad \text{and} \quad C = \begin{bmatrix} a & b \\ c & d \end{bmatrix}$$

If each entry of C is the sum of the corresponding entries of A and B, what are the values for a, b, c, d?

5. Express the following information in matrix form: The number of different models of cars sold during the first quarter of this year is given as:

	January	February	March
Chevette	5	2	2
Escort	2	3	0
Citation	3	1	2
Regal	4	5	4

6. If A is a square matrix and A has 5 rows, what is the number of columns of A?

7. Let

$$A = \begin{bmatrix} 1 & 0 & 3 \\ 2 & -5 & 4 \\ 3 & 2 & 1 \end{bmatrix}$$

Find all of the elements of the main diagonal of A.

8. List all the diagonal elements of the Matrix $B = [b_{ij}]_{n \times n}$

9. If A is a symmetric matrix and the (2,3)-entry of A is a_{23}, what is the (3,2)-entry of A.

10. Find an example of a matrix which is both symmetric and upper triangular.

11. Give an example of a matrix which is both upper and lower triangle, but not a scalar.

12. If A is a lower triangular matrix, what is the (2,3)-entry of A?

13. Let A = [1   2   3].   Find the column matrix (vector) with the same elements as A.

14. Give the 2x2, 3x3, 4x4 identity matrices

15. If A is a scalar matrix, and the (3,3)-entry of A is 5, what are the (4,4)-entry and (3,4)-entry of A?

2.2 MATRIX OPERATIONS

OBJECTIVES:

1. Define equality of matrices.
2. Define the basic operation of addition, scalar multiplication, transposition and multiplication of matrices.
3. Discuss some basic properties of matrix operations.

DEFINITION 2.2: Two matrices are said to be equal if they have the same order and their corresponding elements are equal. Thus, if $A = [a_{ij}]_{mxn}$, $B = [b_{ij}]_{rxs}$, then $A = B$ if and only if $m = r$, $n = s$ and $a_{ij} = b_{ij}$ for all i, j.

For example,

$$\begin{bmatrix} 1 & 2 \\ 3 & -1 \end{bmatrix} = \begin{bmatrix} 1 & \sqrt{4} \\ \frac{9}{3} & -1 \end{bmatrix} , \quad \begin{bmatrix} 3 \\ 3 \end{bmatrix} \neq [3 \quad 2]$$

and if $\begin{bmatrix} x \\ y-1 \end{bmatrix} = \begin{bmatrix} -1 \\ 3 \end{bmatrix}$

then $x = -1$ and $y - 1 = 3$
or $x = -1$ and $y = 4$

SELF-TEST:

(a)   If $A = \begin{bmatrix} x & 3 \\ 0 & y+1 \end{bmatrix}$, $B = \begin{bmatrix} 4 & 3 \\ 0 & -3 \end{bmatrix}$ and $A = B$, then

x = _____ and y = _____.

78

(b)   If $A = \begin{bmatrix} 3 \\ y-1 \end{bmatrix}$, $B = \begin{bmatrix} 3 \\ 5 \end{bmatrix}$,   and $A \neq B$, then y = _____.

ANS:

(a)   x = 4, y = -4

(b)   y = anything except 6

BASIC OPERATIONS OF MATRICES:

1.   Let $A = [a_{ij}]_{m \times n}$, $B = [b_{ij}]_{m \times n}$.   Then the sum of A and B, denoted by A + B, is defined as

$$A + B = [c_{ij}]_{m \times n}$$

where $c_{ij} = a_{ij} + b_{ij}$

Thus, the (i,j)-entry of the sum   of the two matrices A and B is the sum of the corresponding (i,j)-entries of A and B.   Note that we can only add two matrices of the same order.   For example,

$$\text{if } A = \begin{bmatrix} 1 & 4 \\ 2 & 3 \end{bmatrix}, \quad B = \begin{bmatrix} 2 & 5 \\ 0 & 4 \end{bmatrix}$$

$$\text{then } A + B = \begin{bmatrix} 1+2 & 4+5 \\ 2+0 & 3+4 \end{bmatrix} = \begin{bmatrix} 3 & 9 \\ 2 & 7 \end{bmatrix}$$

SELF-TEST:

(a)   If $C = \begin{bmatrix} 3 & -1 & 0 \\ 2 & 4 & 1 \end{bmatrix}$, $D = \begin{bmatrix} 7 & 3 & 5 \\ -1 & 3 & -2 \end{bmatrix}$,

then C + D = $\begin{bmatrix} \\ \end{bmatrix}$

(b) If $A = \begin{bmatrix} 3 & x \\ 5 & -1 \end{bmatrix}$, $B = \begin{bmatrix} -2 & 4 \\ y & 3 \end{bmatrix}$, and $A + B = \begin{bmatrix} 1 & 6 \\ 3 & 2 \end{bmatrix}$,

then x = _____ , y = _____ .

ANS:

(a)
$$C + D = \begin{bmatrix} 10 & 2 & 5 \\ 1 & 7 & -1 \end{bmatrix}$$

(b) x + 4 = 6, hence x = 2
     y + 5 = 3, hence y = -2

2. <u>Scalar multiplication:</u> If $A = [a_{ij}]_{m \times n}$, and k is a real number (scalar), then the scalar multiple of A by k, denoted by kA, is defined as $kA = [ka_{ij}]_{m \times n}$. That is, the (i,j)-entry of the scalar multiple is obtained by multiplying the (i,j)-entry of A by k.

For example, if $A = \begin{bmatrix} 4 & -1 \\ 3 & 2 \end{bmatrix}$ and k = 5, then

$$5A = \begin{bmatrix} 5 \times 4 & 5 \times (-1) \\ 5 \times 3 & 5 \times 2 \end{bmatrix} = \begin{bmatrix} 20 & -5 \\ 15 & 10 \end{bmatrix}$$

Also, if $B = \begin{bmatrix} -3 & 2 & 4 \\ 0 & 3 & 1 \end{bmatrix}$, k = -1, then

$$-1B = \begin{bmatrix} 3 & -2 & -4 \\ 0 & -3 & -1 \end{bmatrix} = -B.$$

Note that we write -1B as -B. It is clear that the zero multiple of any matrix is the zero matrix, i.e., the matrix with every entry zero.

(a) If $A = \begin{bmatrix} 3 & \frac{1}{2} \\ -1 & 4 \end{bmatrix}$, and $k = 6$ then $k \cdot A = \begin{bmatrix} & \\ & \end{bmatrix}$

(b) If $B = \begin{bmatrix} 5 & 0 & -1 \\ 4 & x & 3 \end{bmatrix}$, and $C = 2$ and $CB = \begin{bmatrix} 10 & 0 & -2 \\ 8 & -6 & 6 \end{bmatrix}$

then $x = \underline{\hspace{2cm}}$.

ANS:

(a) $k \cdot A = \begin{bmatrix} 18 & 3 \\ -6 & 24 \end{bmatrix}$

(b) $2x = -6$, hence $x = -3$

3. Transposition: Let $A = [a_{ij}]_{m \times n}$ be a matrix with m rows and n columns whose (i,j)-entry is a_{ij}. The transpose of A, denoted by A', is defined as

$$A' = [b_{ij}]_{n \times m}$$

where $b_{ij} = a_{ji}$, i.e., the (i,j)-entry of the transpose of A is the (j,i)-entry of the matrix A. Note that the transpose A' has n rows and m columns. For example, if

$$A = \begin{bmatrix} 1 & 2 & 3 \\ 4 & 5 & 6 \end{bmatrix}$$

then

$$A' = \begin{bmatrix} 1 & 4 \\ 2 & 5 \\ 3 & 6 \end{bmatrix}.$$

Notice that each row of A becomes a column of A' and vice versa. If A is a symmetric matrix, or each (i,j)-entry

81

of A is equal to the (j, i) -entry of A, then A and its transpose A' coincide. For example, if

$$A = \begin{bmatrix} 1 & -3 \\ -3 & 4 \end{bmatrix}, \quad \text{then } A' = \begin{bmatrix} 1 & -3 \\ -3 & 4 \end{bmatrix} = A.$$

Therefore, it is clear that

$$A \text{ is a symmetric matrix} \quad \underset{\text{and only if}}{\overset{\text{if}}{\Longleftrightarrow}} \quad A = A'.$$

SELF-TEST:

(a) $A = \begin{bmatrix} 3 & -1 \\ 2 & 0 \\ 4 & 9 \end{bmatrix}$ $\qquad A' = \begin{bmatrix} & \\ & \end{bmatrix}$

(b) If $B' = \begin{bmatrix} 3 & 2 \\ y & 4 \end{bmatrix}$ and if B is symmetric, then y = _____

ANS:

(a) $A' = \begin{bmatrix} 3 & 2 & 4 \\ -1 & 0 & 9 \end{bmatrix}$

(b) $B = \begin{bmatrix} 3 & y \\ 2 & 4 \end{bmatrix}$ so y = 2

4. <u>Matrix multiplication</u>: First a special case:

Let $A = [a_1 \; a_2 \; \ldots \; a_n]$ be a row matrix with n entries

and $B = \begin{bmatrix} b_1 \\ b_2 \\ \vdots \\ b_n \end{bmatrix}$ be a column matrix with also n entries. The

product A and B, denoted by A·B, is defined as

$$A \cdot B = [a_1 b_1 + a_2 b_2 + \ldots + a_n b_n]$$

82

Note that in this case the product A·B has only one entry.

For example, if $A = [1 \quad 4]$, $B = \begin{bmatrix} 2 \\ 5 \end{bmatrix}$ then $A·B = [1·2 + 4·5]$

$= [22]$.

Another example, if $C = [1 \quad 4 \quad -2]$, $D = \begin{bmatrix} 3 \\ -1 \\ 5 \end{bmatrix}$ then

$C·D = [1·3 + 4·(-1) + (-2)·5] = [-11]$.

SELF-TEST:

(a)   If $A = [2 \quad 0 \quad 4]$, $B = \begin{bmatrix} -1 \\ 9 \\ 3 \end{bmatrix}$, then $A·B = [\quad]$.

(b)   If $A = [x \quad y]$, $B = \begin{bmatrix} 3 \\ -9 \end{bmatrix}$, then $A·B = [\quad]$.

ANS: (a) $A·B = [10]$

(b) $A·B = [3x-9y]$

Now the general case: Let $A = [a_{ij}]_{mxn}$, $B = [b_{ij}]_{nxp}$. The product of A and B, denoted by A·B, is defined as

$$A·B = C = [c_{ij}]_{mxp}$$

where $c_{ij} = a_{i1}b_{1j} + a_{i2}b_{2j} + \ldots + a_{in}b_{nj}$, i.e., the (i,j)-entry of the product A·B is obtained by multiplying the i-th row of the matrix A and the j-th column of the matrix B which can be considered as multiplying a row matrix by a column matrix:

$$A = \begin{bmatrix} a_{11} & a_{12} \ldots a_{1n} \\ a_{21} & a_{22} \ldots a_{2n} \\ \boxed{a_{i1}} & a_{i2} \ldots a_{in} \\ a_{m1} & a_{m2} \ldots a_{mn} \end{bmatrix} \leftarrow i\text{-th row},$$

$$B = \begin{bmatrix} b_{11} & b_{12} & \cdots & \boxed{b_{1j}} & \cdots & b_{1p} \\ b_{21} & b_{22} & \cdots & b_{2j} & \cdots & b_{2p} \\ \vdots & \vdots & & \vdots & & \vdots \\ b_{n1} & b_{n2} & \cdots & \boxed{b_{nj}} & \cdots & b_{np} \end{bmatrix}$$

$$\uparrow$$
$$j\text{-th column}$$

$$A \cdot B = \begin{bmatrix} \cdots & \cdots & \cdots & \cdots & \cdots & \cdots \\ \cdots & \cdots & \cdots & \cdots & \cdots & \cdots \\ \cdots & \boxed{a_{i1}b_{1j} + a_{i2}b_{2j} + \cdots + a_{in}b_{nj}} & \cdots \\ \cdots & \cdots & \cdots & \cdots & \cdots & \cdots \end{bmatrix} \leftarrow \begin{array}{l}(ij)\text{-entry} \\ \text{of } A \cdot B.\end{array}$$

For example, if $A = \begin{bmatrix} 1 & 4 \\ -2 & 3 \end{bmatrix}$, $B = \begin{bmatrix} -1 & 3 & 4 \\ 2 & 5 & -2 \end{bmatrix}$, then

$$A \cdot B = \begin{bmatrix} 1(-1) + 4 \cdot 2 & 1 \cdot 3 + 4 \cdot 5 & 1 \cdot 4 + 4(-2) \\ (-2)(-1) + 3 \cdot 2 & (-2)3 + 3 \cdot 5 & (-2)4 + 3(-2) \end{bmatrix} = \begin{bmatrix} 7 & 23 & -4 \\ 8 & 9 & -14 \end{bmatrix}$$

NOTE:

1. The product A·B is defined only if the number of columns of A is the same as the number of rows of B. For example, if A is a 2x3 matrix and B is a 4x2 matrix, then the product A·B is not defined since A has 3 **columns** and B has 4 rows. However, the product B·A is defined since B has 2 columns which is the same as the number of rows in A.

2. If A is an $m \times n$ matrix and B an $n \times p$ matrix, then the product A·B is a $m \times p$ matrix, i.e., the product A·B is a matrix with the same number of rows as A and same number of columns as B.

From the remark made above, we know that it is possible that given two matrices A and B, the product A·B is defined while the product B·A is not defined. Thus, matrix multipliction is clearly **not** commutative, i.e., in general,

84

A·B is not equal to B·A.

Even in the case that both A·B and B·A are defined, they are still not necessarily equal to each other as can be seen from the following example:

Let $A = \begin{bmatrix} 1 & 3 \\ -2 & 4 \end{bmatrix}$, $B = \begin{bmatrix} -1 & 1 \\ 2 & 3 \end{bmatrix}$.

$A \cdot B = \begin{bmatrix} 5 & 10 \\ 10 & 10 \end{bmatrix}$

$B \cdot A = \begin{bmatrix} -3 & 1 \\ -4 & 18 \end{bmatrix}$

$A \cdot B \neq B \cdot A$

Therefore, we know that matrix multiplication is not nec- essarily commutative.

SELF-TEST:

(a) If $A = \begin{bmatrix} 2 & -1 & 0 \\ -3 & 4 & 5 \end{bmatrix}$, $B = \begin{bmatrix} 1 & 3 \\ -1 & 2 \\ 4 & -5 \end{bmatrix}$,

$A \cdot B = \begin{bmatrix} & \\ & \end{bmatrix}$, $B \cdot A = \begin{bmatrix} & \\ & \end{bmatrix}$,

and the product A·B has _____ rows and _____ columns. The product B·A has _____ rows and _____ columns.

(b) If C is a 5x3 matrix, and D is a 7x5 matrix, why is the product C·D not defined? How about the matrix D·C? What is the size of the product D·C?

(c) If $A = \begin{bmatrix} x & 3 \\ 2 & 1 \end{bmatrix}$, $B = \begin{bmatrix} -2 & 0 \\ 2 & 4 \end{bmatrix}$ and $A \cdot B = \begin{bmatrix} 4 & 12 \\ y & z \end{bmatrix}$

85

then x = _____ , y = _____ , z = _____ .

ANS:

(a) $A \cdot B = \begin{bmatrix} 3 & 4 \\ 13 & -26 \end{bmatrix}$, $B \cdot A = \begin{bmatrix} -7 & 11 & 15 \\ -8 & 9 & 10 \\ 23 & -24 & -25 \end{bmatrix}$

A·B has 2 rows and 2 columns.

B·A has 3 rows and 3 columns.

(b)   C·D is not defined because the number of columns of C is 3 which is not equal to the numbers of rows of D which is 7. D·C is a 7 x 3 matrix.

(c)   -2x + 3·2 = 4, hence x = 1

y = 2(-2) + 1·2 = -2

z = 2·0 + 1·4 = 4

Some basic properties of matrix operations: Assuming all sums and products are defined, then the following properties for matrix operations hold.

1.   Associative property for multiplication: $(A \cdot B) \cdot C = A \cdot (B \cdot C)$
     For example, if

$$A = \begin{bmatrix} 2 & -1 \\ 1 & 0 \end{bmatrix}, \quad B = \begin{bmatrix} 1 & -1 \\ 2 & 3 \end{bmatrix}, \quad C = \begin{bmatrix} 4 \\ -2 \end{bmatrix},$$

then $(A \cdot B) \cdot C = \left(\begin{bmatrix} 2 & -1 \\ 1 & 0 \end{bmatrix} \cdot \begin{bmatrix} 1 & -1 \\ 2 & 3 \end{bmatrix} \right) \cdot \begin{bmatrix} 4 \\ -2 \end{bmatrix}$

$$= \begin{bmatrix} 0 & -5 \\ 1 & -1 \end{bmatrix} \begin{bmatrix} 4 \\ -2 \end{bmatrix} = \begin{bmatrix} 10 \\ 6 \end{bmatrix}$$

and $A \cdot (B \cdot C) = \begin{bmatrix} 2 & -1 \\ 1 & 0 \end{bmatrix} \cdot \left(\begin{bmatrix} 1 & -1 \\ 2 & 3 \end{bmatrix} \cdot \begin{bmatrix} 4 \\ -2 \end{bmatrix} \right)$

86

$$= \begin{bmatrix} 2 & -1 \\ 1 & 0 \end{bmatrix} \cdot \begin{bmatrix} 6 \\ 2 \end{bmatrix} = \begin{bmatrix} 10 \\ 6 \end{bmatrix}$$

2. Distributive properties:

$A \cdot (B+C) = A \cdot B + A \cdot C$

$(A+B) \cdot C = A \cdot C + B \cdot C$

For example, if

$$A = \begin{bmatrix} 0 & 1 \\ 3 & -2 \\ -1 & 1 \end{bmatrix}, \quad B = \begin{bmatrix} 4 & 2 \\ -1 & 1 \end{bmatrix}, \quad C = \begin{bmatrix} -3 & 5 \\ 2 & 4 \end{bmatrix}$$

$$A \cdot (B+C) = \begin{bmatrix} 0 & 1 \\ 3 & -2 \\ -1 & 1 \end{bmatrix} \left(\begin{bmatrix} 4 & 2 \\ -1 & 1 \end{bmatrix} + \begin{bmatrix} -3 & 5 \\ 2 & 4 \end{bmatrix} \right)$$

$$= \begin{bmatrix} 0 & 1 \\ 3 & -2 \\ -1 & 1 \end{bmatrix} \begin{bmatrix} 1 & 7 \\ 1 & 5 \end{bmatrix} = \begin{bmatrix} 1 & 5 \\ 1 & 11 \\ 0 & -2 \end{bmatrix}$$

$$A \cdot B + A \cdot C = \begin{bmatrix} 0 & 1 \\ 3 & -2 \\ -1 & 1 \end{bmatrix} \begin{bmatrix} 4 & 2 \\ -1 & 1 \end{bmatrix} + \begin{bmatrix} 0 & 1 \\ 3 & -2 \\ -1 & 1 \end{bmatrix} \begin{bmatrix} -3 & 5 \\ 2 & 4 \end{bmatrix}$$

$$= \begin{bmatrix} -1 & 1 \\ 14 & 4 \\ -5 & -1 \end{bmatrix} + \begin{bmatrix} 2 & 4 \\ -13 & 7 \\ 5 & -1 \end{bmatrix} = \begin{bmatrix} 1 & 5 \\ 1 & 11 \\ 0 & -2 \end{bmatrix}$$

3. If I is the identity matrix, then

$I \cdot A = A$

$A \cdot I = A$

For example, if

$$I = \begin{bmatrix} 1 & 0 \\ 0 & 1 \end{bmatrix}, \quad A = \begin{bmatrix} 4 & 5 \\ -3 & 2 \end{bmatrix}$$

$$I \cdot A = \begin{bmatrix} 4 & 5 \\ -3 & 2 \end{bmatrix}$$

$$A \cdot I = \begin{bmatrix} 4 & 5 \\ -3 & 2 \end{bmatrix}$$

SELF-TEST:

Let $A = \begin{bmatrix} 5 & 0 \\ 3 & -2 \end{bmatrix}$, $B = \begin{bmatrix} 1 & -3 \\ 4 & 2 \end{bmatrix}$, $C = \begin{bmatrix} -1 & 1 \\ 3 & -2 \end{bmatrix}$

(a) compute $(A+B) \cdot C$ and $A \cdot C + B \cdot C$.

(b) Compute $(A \cdot B) \cdot C$ and $A \cdot (B \cdot C)$

(c) If $I = \begin{bmatrix} 1 & 0 \\ 0 & 1 \end{bmatrix}$, compute $I \cdot A$ and $C \cdot I$.

ANS:

(a) $(A+B) \cdot C = A \cdot C + B \cdot C = \begin{bmatrix} -15 & 12 \\ -7 & 7 \end{bmatrix}$

(b) $(A \cdot B) \cdot C = A \cdot (B \cdot C) = \begin{bmatrix} -50 & 35 \\ -34 & 21 \end{bmatrix}$

(c) $I \cdot A = \begin{bmatrix} 5 & 0 \\ 3 & -2 \end{bmatrix}$ $\quad C \cdot I = \begin{bmatrix} -1 & 1 \\ 3 & -2 \end{bmatrix}$

EXERCISES 2.2

1. Let $A = \begin{bmatrix} 4 & 2 \\ -1 & x \end{bmatrix}$, $B = \begin{bmatrix} 4 & y \\ -1 & 7 \end{bmatrix}$

    If A = B, then x = _____? y = _____?

2. Let $C = \begin{bmatrix} 4+x & 5+y \\ 3 & 4 \end{bmatrix}$, $D = \begin{bmatrix} -1 & 0 \\ 3 & z \end{bmatrix}$

    If C = D, what are the values for x, y, z?

3. If $A = [x^2-1 \quad 4]$, $B = [3 \quad 4]$ and if A = B, then
    x = _____?

4. Let $A = \begin{bmatrix} 2 & 3 \\ -1 & 0 \end{bmatrix}$, $B = \begin{bmatrix} 1 & 2 \\ -1 & 3 \\ 0 & 4 \end{bmatrix}$, $C = \begin{bmatrix} -1 & 0 & 1 \\ 1 & 1 & 2 \end{bmatrix}$

    compute each of the following if possible:
    (a) A + C
    (b) C + A
    (c) B + C
    (d) A + BC
    (e) (A+C)·B
    (f) B·(A+C)
    (g) B·A + B·C
    (h) (A·C)·B
    (i) A·(C·B)
    (j) (B·A)·C

5. Let $A = \begin{bmatrix} 3 & -1 \\ 2 & 6 \\ 7 & 9 \end{bmatrix}$ Compute

    (a) 3A
    (b) −A
    (c) (−1)·A
    (d) A'

89

6. If $B = \begin{bmatrix} 2 & x \\ -1 & 3 \end{bmatrix}$ and B is symmetric, then x = _____?

7. Give an example of two square matrices A, B such that A·B ≠ B·A.

8. If A is 3x5, B is 5x2, C is 2x7, then the order of the matrix (A·B)·C is _____?

9. If $A = \begin{bmatrix} 4 & 5 & 6 \\ 3 & -1 & 0 \end{bmatrix}$, $I = \begin{bmatrix} 1 & 0 & 0 \\ 0 & 1 & 0 \\ 0 & 0 & 1 \end{bmatrix}$ then A·I = _____

   How about I·A?

10. Let $A = \begin{bmatrix} 1 & 2 \\ -1 & 5 \end{bmatrix}$, $B = \begin{bmatrix} 0 & 3 \\ 4 & 2 \end{bmatrix}$ Compute

   (a) (A+B)'
   (b) A' + B'
   (c) (A·B)'
   (d) A' · B'
   (e) B' • A'

11. Same as 10, but $A = \begin{bmatrix} 1 & -1 & 2 \\ 0 & 3 & 5 \\ 4 & 0 & -2 \end{bmatrix}$, $B = \begin{bmatrix} 1 & -2 & 3 \\ 4 & 7 & 5 \\ 1 & 1 & 0 \end{bmatrix}$

12. Let $A = \begin{bmatrix} 4 & -1 \\ 1 & 3 \end{bmatrix}$, $B = \begin{bmatrix} 2 & 5 \\ -1 & 0 \end{bmatrix}$, a = 3, d = 5. Compute

   (a) a·(A+B)
   (b) a·A + a·B
   (c) (a+d)·A
   (d) a·A + d·A

13. Let $A = \begin{bmatrix} 1 & 2 & 3 \\ 0 & 0 & 0 \\ -1 & 1 & 2 \end{bmatrix}$, $B = \begin{bmatrix} 1 & 1 \\ -1 & 0 \\ 2 & 4 \end{bmatrix}$. Compute A·B.

14. Let $A = \begin{bmatrix} 1 & -1 & 1 \\ 2 & 4 & 0 \\ 3 & -1 & 2 \end{bmatrix}$, $B = \begin{bmatrix} 1 & 1 & 0 \\ 3 & 2 & 0 \\ -1 & -2 & 0 \end{bmatrix}$. Compute $A \cdot B$.

15. If $A = [a_{ij}]_{m \times n}$, $B = [b_{ij}]_{n \times p}$ and

    (a) if the k=th row of A has all zero elements, what can you say about $A \cdot B$?

    (b) if the k-th column of B has all zero elements, what can you say about $A \cdot B$?

2.3 Row Reduced Form of a Matrix and Matrix Inversion

Elementary row operations:

There are three elementary row operations:

I. Interchanging two rows of a matrix. For example,
if $A = \begin{bmatrix} 1 & 3 \\ -4 & 5 \\ 0 & 2 \end{bmatrix}$, interchange the 2nd row and the
3rd row of A and we obtain the matrix $B = \begin{bmatrix} 1 & 3 \\ 0 & 2 \\ -4 & 5 \end{bmatrix}$

II. Multiplying a row of a matrix by a nonzero constant.

For example, if $A = \begin{bmatrix} 2 & 3 \\ -5 & 4 \end{bmatrix}$, multiple the 2nd row
of A by 2 and we obtain the matrix $B = \begin{bmatrix} 2 & 3 \\ -10 & 8 \end{bmatrix}$.

III. Adding a multiple of one row of a matrix to another
row. For example, if $A = \begin{bmatrix} 1 & 2 & -3 \\ 4 & -1 & 5 \\ 0 & 3 & 2 \end{bmatrix}$, adding 2 times
the second row of A to the 3rd row of A, we obtain

$$\text{the matrix } B = \begin{bmatrix} 1 & 2 & -3 \\ 4 & -1 & 5 \\ 0+2(4) & 3+2(-1) & 2+2(5) \end{bmatrix} = \begin{bmatrix} 1 & 2 & -3 \\ 4 & -1 & 5 \\ 8 & 1 & 12 \end{bmatrix}$$

SELF-TEST:

Let $A = \begin{bmatrix} 3 & 4 & -2 & 1 \\ 7 & 0 & 1 & 5 \\ 2 & -9 & 3 & 2 \end{bmatrix}$, $B = \begin{bmatrix} 4 & 3 \\ -2 & 0 \end{bmatrix}$

(a)   Find the matrix C which is obtained from A by interchanging the first row and second row of A.

$$C = \begin{bmatrix} & & \\ & & \end{bmatrix}$$

(b)   Find the matrix D which is obtained from B by multiplying the first row of B by -5.

$$D = \begin{bmatrix} & \\ & \end{bmatrix}$$

(c)   Find the matrix E which is obtained from A by adding -2 times the first row of A to the second row of A.

$$E = \begin{bmatrix} & & \\ & & \end{bmatrix}$$

ANS: (a) $C = \begin{bmatrix} 7 & 0 & 1 & 5 \\ 3 & 4 & -2 & 1 \\ 2 & -9 & 3 & 2 \end{bmatrix}$

(b) $D = \begin{bmatrix} -20 & -15 \\ -2 & 0 \end{bmatrix}$

(c) $E = \begin{bmatrix} 3 & 4 & -2 & 1 \\ 1 & -8 & 5 & 3 \\ 2 & -9 & 3 & 2 \end{bmatrix}$

Row reduced form of a matrix:

Let $A = [a_{ij}]_{m \times n}$. A is said to be in reduced form if the following conditions are satisfied.

1. The first non-zero element of each row of A must be 1;
2. if the (i,j)-entry is the first non-zero element of the i-th row and the (r,s)-entry is that of the r-th row, and $i < r$, then j must be less than s. As an example, suppose the first row of the matrix A is 0 1 2 4, i.e., the $(1,2)$-entry is the first non-zero element of row 1 of A, then in row 2, the first non-zero element must occur in column 3 or column 4;
3. all entries lying below or above the first non-zero element of each row must be zero; for example, if the $(2,3)$-entry is the first non-zero element of row 2 of the matrix A, then all other elements in column 3 of A must be zero.

NOTE: Some rows of a row-reduced matrix may be zero, but they must stay below the non-zero rows.

Examples of row-reduced matrices:

$$\begin{bmatrix} 1 & 0 & 1 \\ 0 & 1 & 3 \\ 0 & 0 & 0 \end{bmatrix} \quad \begin{bmatrix} 1 & 2 & 0 \\ 0 & 0 & 1 \end{bmatrix} \quad \begin{bmatrix} 1 & 0 & 0 \\ 0 & 1 & 0 \\ 0 & 0 & 1 \end{bmatrix}$$

$$\begin{bmatrix} 0 & 1 & 0 & 3 \\ 0 & 0 & 1 & 1 \\ 0 & 0 & 0 & 0 \\ 0 & 0 & 0 & 0 \end{bmatrix}$$

The following matrices are not in row-reduced form:

$$\begin{bmatrix} 1 & 0 & 3 \\ 0 & 2 & 0 \\ 0 & 0 & 0 \end{bmatrix}$$
The first non-zero element of row 2 is not 1.

$$\begin{bmatrix} 0 & 1 & 0 \\ 1 & 0 & 0 \end{bmatrix}$$
The first non-zero element of row 1 is in column 2, but the first non-zero element of row 1 is in column 1.

$$\begin{bmatrix} 0 & 1 & 2 \\ 0 & 0 & 1 \end{bmatrix}$$
The (1,3)-entry is not zero

$$\begin{bmatrix} 1 & 0 & 0 \\ 0 & 1 & 0 \\ 0 & 1 & 0 \end{bmatrix}$$
The (3,2)-entry is not zero.

SELF-TEST:

Determine which of the following matrices are in row-reduced from:

(a) $\begin{bmatrix} 1 & 2 & 0 \\ 0 & 0 & 1 \end{bmatrix}$ (b) $\begin{bmatrix} 0 & 1 & -1 & 2 \\ 0 & 0 & 0 & 1 \end{bmatrix}$

(c) $\begin{bmatrix} 0 & 0 & 1 & 0 \\ 0 & 0 & 0 & -1 \end{bmatrix}$ (d) $\begin{bmatrix} 1 & 4 & 0 & 2 & 0 \\ 0 & 0 & 1 & 5 & 0 \\ 0 & 0 & 0 & 0 & 1 \end{bmatrix}$

ANS: a, d are in row-reduced form.

Given any matrix A, we can change it into row-reduced form by repeatedly applying the three elementary row operations.

For example, let A = $\begin{bmatrix} 2 & 3 \\ 1 & 2 \end{bmatrix}$. Interchange the first and second rows of A and we obtain $A_1 = \begin{bmatrix} 1 & 2 \\ 2 & 3 \end{bmatrix}$. Add -2 times the first row to the second row of the matrix A_1 we obtain $A_2 = \begin{bmatrix} 1 & 2 \\ 0 & -1 \end{bmatrix}$.

Multiply the second row of A_2 by -1 and we obtain $A_3 = \begin{bmatrix} 1 & 2 \\ 0 & 1 \end{bmatrix}$.

Now, add -2 times the second row to the first row of A_3 and we o

$A_4 = \begin{bmatrix} 1 & 0 \\ 0 & 1 \end{bmatrix}$ and A_4 is in row-reduced form,

As another example, let $B = \begin{bmatrix} 3 & 0 & -6 & 3/2 \\ 0 & 0 & -2 & 4 \\ -2 & 0 & 4 & -1 \end{bmatrix}$. To change

B to row-reduced form, we first change the first non-zero elemen of row 1 of B into 1. This can be done by multiplying each of the elements of row 1 by $1/3$, thus we obtain

$$B_1 = \begin{bmatrix} 1 & 0 & -2 & \tfrac{1}{2} \\ 0 & 0 & -2 & 4 \\ -2 & 0 & 4 & -1 \end{bmatrix}$$

To make every element below the $(1,1)$-entry of B_1 a zero, we add 2 times row 1 to row 3 and obtain:

$$B_2 = \begin{bmatrix} 1 & 0 & -2 & \tfrac{1}{2} \\ 0 & 0 & -2 & 4 \\ 0 & 0 & 0 & 0 \end{bmatrix}$$

To change the first non-zero element of row 2 of B_2 to a 1, we multiply the second row by $-1/2$ and obtain:

$$B_3 = \begin{bmatrix} 1 & 0 & -2 & \tfrac{1}{2} \\ 0 & 0 & 1 & -2 \\ 0 & 0 & 0 & 0 \end{bmatrix}$$

To change B_3 to row-reduced form, we only have to add 2 times row 2 of B_3 to row 1 and the matrix

$$B_4 = \begin{bmatrix} 1 & 0 & 0 & -\frac{7}{2} \\ 0 & 0 & 1 & -2 \\ 0 & 0 & 0 & 0 \end{bmatrix}$$

is indeed in row-reduced form.

One more example:

$$A = \begin{bmatrix} 2 & 1 & 0 \\ 0 & -2 & 3 \\ 1 & 4 & 5 \end{bmatrix} \xrightarrow[\text{and third row}]{\substack{\text{interchange} \\ \text{the first}}} \begin{bmatrix} 1 & 4 & 5 \\ 0 & -2 & 3 \\ 2 & 1 & 0 \end{bmatrix}$$

$$\xrightarrow{\text{row } 3 + (-2) \cdot \text{row } 1} \begin{bmatrix} 1 & 4 & 5 \\ 0 & -2 & 3 \\ 0 & -7 & -10 \end{bmatrix}$$

$$\xrightarrow{\text{row } 3 + (-4) \cdot \text{row } 2} \begin{bmatrix} 1 & 4 & 5 \\ 0 & -2 & 3 \\ 0 & 1 & -22 \end{bmatrix}$$

$$\xrightarrow[\text{and row } 3]{\text{interchange row } 2} \begin{bmatrix} 1 & 4 & 5 \\ 0 & 1 & -22 \\ 0 & -2 & 3 \end{bmatrix}$$

$$\xrightarrow{\text{row } 3 + 2 \cdot \text{row } 2} \begin{bmatrix} 1 & 4 & 5 \\ 0 & 1 & -22 \\ 0 & 0 & -41 \end{bmatrix}$$

$$\xrightarrow[\text{by } -1/41]{\text{multiple row } 3} \begin{bmatrix} 1 & 4 & 5 \\ 0 & 1 & -22 \\ 0 & 0 & 1 \end{bmatrix}$$

$$\xrightarrow{\text{row } 2 + 22 \cdot \text{row } 3} \begin{bmatrix} 1 & 4 & 5 \\ 0 & 1 & 0 \\ 0 & 0 & 1 \end{bmatrix}$$

$$\xrightarrow{\text{row } 1 + (-5) \cdot \text{row } 3} \quad \begin{bmatrix} 1 & 4 & 0 \\ 0 & 1 & 0 \\ 0 & 0 & 1 \end{bmatrix}$$

$$\xrightarrow{\text{row } 1 + (-4) \cdot \text{row } 2} \quad \begin{bmatrix} 1 & 0 & 0 \\ 0 & 1 & 0 \\ 0 & 0 & 1 \end{bmatrix}$$

The last matrix is in row-reduced form.

REMARKS:

1. If A is a square matrix and A is in row-reduced form, then either A has at least a zero row, i.e., some row of A has only zero elements, or A is the indentity matrix.

2. If A has more rows than columns and if A is in row-reduced form, then A must have at least one zero row.

3. We will see in the next section that the row-reduced form of a matrix will help us to find the inverse of a matrix if it exists. Also, in section 2.4 we will learn how to solve a system of linear equations, by reducing a matrix to row-reduced form.

SELF-TEST:

Change each of the following matrices to row-reduced form:

(a) $\begin{bmatrix} 5 & 4 \\ 3 & 6 \end{bmatrix}$
 (b) $\begin{bmatrix} 2 & 4 & -2 \\ 3 & 5 & 1 \\ 5 & 9 & -1 \end{bmatrix}$
 (c) $\begin{bmatrix} 0 & 4 & 0 & 3 \\ 0 & -2 & 0 & -3 \\ 0 & 2 & 0 & 0 \end{bmatrix}$

(a) $\begin{bmatrix} 1 & 0 \\ 0 & 1 \end{bmatrix}$

(b) $\begin{bmatrix} 1 & 0 & 7 \\ 0 & 1 & -4 \\ 0 & 0 & 0 \end{bmatrix}$

(c) $\begin{bmatrix} 0 & 1 & 0 & 0 \\ 0 & 0 & 0 & 1 \\ 0 & 0 & 0 & 0 \end{bmatrix}$

Matrix inverting:

DEFINITION 2.3:   A matrix A is said to be invertible if there exists a matrix B such that A·B = B·A = I, where I is the identity matrix.   The matrix B is called the inverse of A.

We state the following facts without proof:

FACT I.   If A is invertible, then A must be a square matrix.

FACT II.   If A is invertible, then its inverse is unique and is denoted by A^{-1}.

FACT III.   Given an invertible matrix A, if there exists a matrix B such that A·B = I (or B·A = I) then B is the inverse of A.

We first consider a special case.   Let

$$A = \begin{bmatrix} a & b \\ c & d \end{bmatrix}$$

be a 2x2 matrix.   The matrix A is invertible if and only if a·d − b·c ≠ 0.

NOTE: ad − bc is called the determinant of A.

If ad − bc ≠ 0, the matrix B defined as

99

$$B = \frac{1}{ad-bc}\begin{bmatrix} d & -b \\ -c & a \end{bmatrix}$$

can be seen as the inverse of A.

Since,

$$B \cdot A = \frac{1}{ad-bc}\begin{bmatrix} d & -b \\ -c & a \end{bmatrix}\begin{bmatrix} a & b \\ c & d \end{bmatrix}$$

$$= \frac{1}{ad-bc}\begin{bmatrix} da-bc & db-bd \\ -ca+ac & -cb+ad \end{bmatrix}$$

$$= \frac{1}{ad-bc}\begin{bmatrix} ad-bc & 0 \\ 0 & ad-bc \end{bmatrix}$$

$$= \begin{bmatrix} 1 & 0 \\ 0 & 1 \end{bmatrix}$$

Also, $A \cdot B = \begin{bmatrix} 1 & 0 \\ 0 & 1 \end{bmatrix}$

Thus, $A \cdot B = B \cdot A = I$ and $B = A^{-1}$.

EXAMPLE. Let $A = \begin{bmatrix} 3 & 1 \\ 2 & 1 \end{bmatrix}$. The matrix A is invertible since

$3 \cdot 1 - 2 \cdot 1 = 1$ is not zero and the inverse is

$$A^{-1} = \frac{1}{1}\begin{bmatrix} 1 & -1 \\ -2 & 3 \end{bmatrix} = \begin{bmatrix} 1 & -1 \\ -2 & 3 \end{bmatrix}$$

CHECK: $A \cdot A^{-1} = \begin{bmatrix} 3 & 1 \\ 2 & 1 \end{bmatrix}\begin{bmatrix} 1 & -1 \\ -2 & 3 \end{bmatrix} = \begin{bmatrix} 1 & 0 \\ 0 & 1 \end{bmatrix}$

EXAMPLE. The matrix $B = \begin{bmatrix} 4 & 2 \\ 2 & 1 \end{bmatrix}$ is not invertible since

$4 \cdot 1 - 2 \cdot 2 = 0$

EXAMPLE. The matrix $C = \begin{bmatrix} x & 1 \\ 4 & 2 \end{bmatrix}$ is invertible if and only

if $2x - 4 \cdot 1 \neq 0$ or $x \neq 2$. If $x \neq 2$,

$$C^{-1} = \frac{1}{2x-4} \begin{bmatrix} 2 & -1 \\ -4 & x \end{bmatrix}$$

SELF-TEST:

(a) If A is invertible then A must be a _____ matrix.

(b) Give an example of a square matrix which is not invertible.

(c) What is the inverse of the identity matrix I_n?

(d) Determine if each of the following 2x2 matrices is invertible. Find the inverse of it is invertible.

1. $A = \begin{bmatrix} -2 & 1 \\ 0 & 3 \end{bmatrix}$

2. $B = \begin{bmatrix} -2 & -3 \\ 4 & 6 \end{bmatrix}$

3. $C = \begin{bmatrix} 3 & 2 \\ 5 & 7 \end{bmatrix}$

4. $D = \begin{bmatrix} 4 & x \\ x & 1 \end{bmatrix}$

(a) square

(b) $\begin{bmatrix} 2 & 2 \\ 2 & 2 \end{bmatrix}$

(c) I_n

(d1) A is invertible, and

$$A^{-1} = \begin{bmatrix} -\frac{1}{2} & \frac{1}{6} \\ 0 & \frac{1}{3} \end{bmatrix}$$

(d2) B is not invertible because the determinant c
B is zero.

(d3) C is invertible because
the determinant of C is

11 and $A^{-1} = \frac{1}{11} \begin{bmatrix} 7 & -2 \\ -5 & \vdots \end{bmatrix}$

(d4) D is invertible if and
only if $x^2 \neq 4$ or $x \neq \pm$
in which case

$$A^{-1} = \frac{1}{4-x^2} \begin{bmatrix} 1 & -x \\ -x & 4 \end{bmatrix}$$

We now give a method to determine if a square matrix is
invertible and to find the inverse if it exists. It is based
on the fact that *a matrix is invertible if and only if it can
be changed into the identity matrix by elementary row operations.*
The **method can be** demonstrated by considering the following
example.

$$\text{Let } A = \begin{bmatrix} 2 & 1 & 0 \\ 7 & 3 & 0 \\ -2 & 0 & 2 \end{bmatrix}$$

To determine if A is invertible and find its inverse, we list the matrix A on one side and the identity matrix I_3 on the other side as follows:

$$[\,A \mid I_3\,]$$

$$= \begin{bmatrix} 2 & 1 & 0 & \vdots & 1 & 0 & 0 \\ 7 & 3 & 0 & \vdots & 0 & 1 & 0 \\ -2 & 0 & 2 & \vdots & 0 & 0 & 1 \end{bmatrix}$$

Apply elementary row operations to change A to row-reduced form. Some operations will be applied to the matrix on the right side. If the row-reduced form of A on left side has a zero row, then A is not invertible. Otherwise, the matrix on the left turns out to be the identity matrix, the matrix A is invertible and the matrix on the right is the inverse. We proceed as follows:

$$\begin{bmatrix} 2 & 1 & 0 & \vdots & 1 & 0 & 0 \\ 7 & 3 & 0 & \vdots & 0 & 1 & 0 \\ -2 & 0 & 2 & \vdots & 0 & 0 & 1 \end{bmatrix} \begin{array}{l} \text{interchange} \\ \text{the 1st and} \\ \text{3rd rows.} \end{array} \begin{bmatrix} -2 & 0 & 2 & \vdots & 0 & 0 & 1 \\ 7 & 3 & 0 & \vdots & 0 & 1 & 0 \\ 2 & 1 & 0 & \vdots & 1 & 0 & 0 \end{bmatrix}$$

$$\begin{array}{l} \text{multiple} \\ \text{row 1 by} \\ -\tfrac{1}{2}. \end{array} \begin{bmatrix} 1 & 0 & -1 & \vdots & 0 & 0 & -\tfrac{1}{2} \\ 7 & 3 & 0 & \vdots & 0 & 1 & 0 \\ 2 & 1 & 0 & \vdots & 1 & 0 & 0 \end{bmatrix}$$

$$\begin{array}{l} \text{row 2 +} \\ (-7)\cdot\text{row} \\ 1. \end{array} \begin{bmatrix} 1 & 0 & -1 & \vdots & 0 & 0 & -\tfrac{1}{2} \\ 0 & 3 & 7 & \vdots & 0 & 1 & \tfrac{7}{2} \\ 2 & 1 & 0 & \vdots & 1 & 0 & 0 \end{bmatrix}$$

$$\begin{array}{l} \text{row 3 +} \\ (-2)\cdot\text{row} \\ 1. \end{array} \begin{bmatrix} 1 & 0 & -1 & \vdots & 0 & 0 & -\tfrac{1}{2} \\ 0 & 3 & 7 & \vdots & 0 & 1 & \tfrac{7}{2} \\ 0 & 1 & 2 & \vdots & 1 & 0 & 1 \end{bmatrix}$$

$$\begin{array}{l} \text{inter-} \\ \text{change} \\ \text{row 2 \&} \\ \text{row 3.} \end{array} \begin{bmatrix} 1 & 0 & -1 & \vdots & 0 & 0 & -\tfrac{1}{2} \\ 0 & 1 & 2 & \vdots & 1 & 0 & 1 \\ 0 & 3 & 7 & \vdots & 0 & 1 & \tfrac{7}{2} \end{bmatrix}$$

row 3 +
(-3) row
2

$$\left[\begin{array}{ccc|ccc} 1 & 0 & -1 & 0 & 0 & -\frac{1}{2} \\ 0 & 1 & 2 & 1 & 0 & 1 \\ 0 & 0 & 1 & -3 & 1 & \frac{1}{2} \end{array}\right]$$

row 1 +
row 3

$$\left[\begin{array}{ccc|ccc} 1 & 0 & 0 & -3 & 1 & 0 \\ 0 & 1 & 2 & 1 & 0 & 1 \\ 0 & 0 & 1 & -3 & 1 & \frac{1}{2} \end{array}\right]$$

row 2 +
(-2) row
3

$$\left[\begin{array}{ccc|ccc} 1 & 0 & 0 & -3 & 1 & 0 \\ 0 & 1 & 0 & 7 & -2 & 0 \\ 0 & 0 & 1 & -3 & 1 & \frac{1}{2} \end{array}\right]$$

The matrix on the left is the identity matrix; hence A is invertible.. The inverse of A is the matrix on the right side; that is,

$$A^{-1} = \left[\begin{array}{ccc} -3 & 1 & 0 \\ 7 & -2 & 0 \\ -3 & 1 & \frac{1}{2} \end{array}\right]$$

<u>CHECK:</u>

$$A \cdot A^{-1} = \left[\begin{array}{ccc} 2 & 1 & 0 \\ 7 & 3 & 0 \\ -2 & 0 & 2 \end{array}\right] \left[\begin{array}{ccc} -3 & 1 & 0 \\ 7 & -2 & 0 \\ -3 & 1 & \frac{1}{2} \end{array}\right] = \left[\begin{array}{ccc} 1 & 0 & 0 \\ 0 & 1 & 0 \\ 0 & 0 & 1 \end{array}\right]$$

<u>EXAMPLE.</u> Let

$$B = \left[\begin{array}{ccc} 1 & 4 & 2 \\ -1 & 2 & 0 \\ 1 & 10 & 4 \end{array}\right]$$

Then,

$$\left[B \vdots I_3\right] = \left[\begin{array}{ccc|ccc} 1 & 4 & 2 & 1 & 0 & 0 \\ -1 & 2 & 0 & 0 & 1 & 0 \\ 1 & 10 & 4 & 0 & 0 & 1 \end{array}\right]$$

$$\rightarrow \begin{bmatrix} 1 & 4 & 2 & \vdots & 1 & 0 & 0 \\ 0 & 6 & 2 & \vdots & 1 & 1 & 0 \\ 1 & 10 & 4 & \vdots & 0 & 0 & 1 \end{bmatrix} \rightarrow \begin{bmatrix} 1 & 4 & 2 & \vdots & 1 & 0 & 0 \\ 0 & 6 & 2 & \vdots & 1 & 1 & 0 \\ 0 & 6 & 2 & \vdots & -1 & 0 & 1 \end{bmatrix}$$

$$\rightarrow \begin{bmatrix} 1 & 4 & 2 & \vdots & 1 & 0 & 0 \\ 0 & 6 & 2 & \vdots & 1 & 1 & 0 \\ 0 & 0 & 0 & \vdots & -2 & -1 & 1 \end{bmatrix}$$

The matrix on the left already has a zero row. Hence, we cannot reach the identity matrix. Therefore the matrix B is not invertible.

We may apply the same method to any $m \times m$ matrix and in particular to 2x2 matrix.

SELF-TEST: Determine if each of the following matrices are invertible and find the inverse if it exists.

(a) $A = \begin{bmatrix} 1 & 3 & -1 \\ 2 & 0 & 4 \\ 1 & 1 & 0 \end{bmatrix}$
(b) $B = \begin{bmatrix} 5 & 3 \\ 3 & 2 \end{bmatrix}$

(c) $C = \begin{bmatrix} 1 & 3 & 2 & 4 \\ -1 & 0 & 1 & 1 \\ 0 & 3 & 3 & 5 \\ 2 & 1 & 1 & -2 \end{bmatrix}$
(d) $D = \begin{bmatrix} 1 & 3 & 3 \\ 1 & 4 & 3 \\ 1 & 3 & 4 \end{bmatrix}$

ANS:

(a) $A^{-1} = \begin{bmatrix} -\frac{2}{3} & -\frac{1}{6} & 2 \\ \frac{2}{3} & \frac{1}{6} & -1 \\ \frac{1}{3} & \frac{1}{3} & -1 \end{bmatrix}$

(b) $B^{-1} = \begin{bmatrix} 2 & -3 \\ -3 & 5 \end{bmatrix}$

(c) C is not invertible

105

$$(d) \quad D^{-1} = \begin{bmatrix} 7 & -3 & -3 \\ -1 & 1 & 0 \\ -1 & 0 & 1 \end{bmatrix}$$

EXERCISES 2.3

1. Let $A = \begin{bmatrix} 1 & 3 & -5 \\ 2 & 1 & 4 \\ 0 & 5 & 6 \end{bmatrix}$,

   (a) Find the matrix B which is obtained from A by interchanging the second and third row of A.

   (b) Find the matrix C which is obtained from A by multiplying the first row of A by 4.

   (c) Find the matrix D which is obtained by adding -3. times the third row to the second row of A.

2. Let $A = \begin{bmatrix} 4 & -1 \\ 2 & 3 \end{bmatrix}$ and B be the matrix which is obtained by interchanging the first and second rows of the 2x2 identity matrix; that is,

   $$B = \begin{bmatrix} 0 & 1 \\ 1 & 0 \end{bmatrix}$$

   (a) Compute $C = B \cdot A$

   (b) Obtain the matrix D obtained by interchanging the first and second rows of A.

   (c) Compare C and D.

3. Let $A = \begin{bmatrix} 1 & -1 & 3 \\ 2 & 0 & 4 \\ 5 & -1 & 1 \end{bmatrix}$, and B the matrix obtained by

   multiplying by -3 each element of the second row of the

3x3 identity matrix; that is

$$B = \begin{bmatrix} 1 & 0 & 0 \\ 0 & -3 & 0 \\ 0 & 0 & 1 \end{bmatrix}$$

(a) Compute C = B·A

(b) Find D obtained by multiplying the second row of A by -3.

(c) Compare C and D

4. Let $A = \begin{bmatrix} 1 & 5 \\ -2 & 3 \end{bmatrix}$ and B be the matrix obtained by adding

-2 times the second row to the first row of the 2x2
identity matrix, that is,

$$B = \begin{bmatrix} 1 & -2 \\ 0 & 1 \end{bmatrix}$$

(a) Compute C = B·A

(b) Find the matrix D obtained by adding -2 times the
     second row of A to the first row.

(c) Compare C and D.

5. Determine which of the following matrices is in row-reduced
    form:

(a) $\begin{bmatrix} 1 & 2 \\ 0 & 0 \end{bmatrix}$ (b) $\begin{bmatrix} 0 & 1 \\ 0 & 0 \end{bmatrix}$ (c) $\begin{bmatrix} 1 & 0 \\ 0 & -1 \end{bmatrix}$

(d) $\begin{bmatrix} 1 & 2 & 5 \\ 0 & 0 & 1 \\ 0 & 0 & 0 \end{bmatrix}$ (e) $\begin{bmatrix} 0 & 1 & 0 \\ 0 & 0 & 1 \\ 0 & 0 & 0 \end{bmatrix}$

(f) $\begin{bmatrix} 1 & 0 & 0 & 0 & 2 \\ 0 & 0 & 1 & 5 & 7 \\ 0 & 0 & 0 & 0 & 0 \\ 0 & 0 & 0 & 0 & 0 \end{bmatrix}$ (g) $\begin{bmatrix} 1 & 0 & 0 \\ 0 & 0 & 1 \\ 0 & 1 & 0 \end{bmatrix}$

107

6. Change each of the following matrices to row-reduced form.

(a) $\begin{bmatrix} 2 & 1 \\ 3 & 5 \end{bmatrix}$ (b) $\begin{bmatrix} 0 & 5 \\ 1 & 3 \end{bmatrix}$

(c) $\begin{bmatrix} 2 & 1 & 3 \\ 1 & 0 & 2 \\ -1 & 1 & 3 \end{bmatrix}$ (d) $\begin{bmatrix} 1 & 1 & 1 & 2 \\ 2 & 1 & -1 & 0 \\ 0 & 0 & 1 & 3 \end{bmatrix}$

(e) $\begin{bmatrix} 0 & 1 & 1 & 2 & 1 \\ 0 & 1 & -1 & -1 & 2 \\ 0 & 0 & 1 & 1 & 3 \\ 0 & 0 & 0 & 0 & 4 \end{bmatrix}$

7. Determine which of the following 2x2 matrices are invertible and find its inverse if it is invertible.

(a) $\begin{bmatrix} 3 & -1 \\ 2 & -1 \end{bmatrix}$ (b) $\begin{bmatrix} 6 & -3 \\ -2 & 1 \end{bmatrix}$ (c) $\begin{bmatrix} 1 & 5 \\ 0 & 9 \end{bmatrix}$

(d) $\begin{bmatrix} 2 & 5 \\ -1 & -3 \end{bmatrix}$

8. Determine the value of x so that the matrix $\begin{bmatrix} 2 & 4 \\ 1 & x \end{bmatrix}$ is invertible.

9. Determine the value for x so that the matrix $\begin{bmatrix} x & 1 \\ 5x-6 & x \end{bmatrix}$ is invertible.

10. Apply the row-reduction algorithm to determine if each of the following matrices are invertible and find the inverse if it exists.

(a) $\begin{bmatrix} 1 & -1 & 1 \\ 2 & 1 & 0 \\ -1 & 0 & 1 \end{bmatrix}$ (b) $\begin{bmatrix} -2 & -3 & 1 \\ 1 & -1 & -1 \\ -1 & -4 & 0 \end{bmatrix}$ (c) $\begin{bmatrix} 1 & 2 & -4 & 5 \\ -2 & -4 & 8 & 1 \\ 1 & 1 & 1 \end{bmatrix}$

(d) $\begin{bmatrix} 1 & -1 & 2 & 1 \\ 1 & 0 & 0 & 1 \\ -1 & 1 & -1 & 1 \\ 2 & 0 & 0 & -1 \end{bmatrix}$

11. An elementary matrix is a matrix obtained by applying an elementary row operation to the identity matrix. It is a fact that an elementary matrix is invertible. Find the inverse of each of the following elementary matrices.

(a) $\begin{bmatrix} 0 & 1 & 0 \\ 1 & 0 & 0 \\ 0 & 0 & 1 \end{bmatrix}$. (b) $\begin{bmatrix} 1 & 0 & 0 & 0 \\ 0 & 2 & 0 & 0 \\ 0 & 0 & 1 & 0 \\ 0 & 0 & 0 & 1 \end{bmatrix}$ (c) $\begin{bmatrix} 1 & 0 & 0 \\ -3 & 1 & 0 \\ 0 & 0 & 1 \end{bmatrix}$

12. Find a matrix $A \neq 0$, (i.e., A is not the matrix consisting of all zeros) such that $A^2 = 0$.
NOTE: $A^2 = A \cdot A$

13. Find two distinct non-zero matrices A, B such at $A \cdot B = 0$, but $B \cdot A \neq 0$.

14. Let A $\begin{bmatrix} 1 & 0 \\ 0 & 0 \end{bmatrix}$, B = $\begin{bmatrix} 0 & 0 \\ 2 & 3 \end{bmatrix}$, C = $\begin{bmatrix} 0 & 0 \\ -1 & -2 \end{bmatrix}$, $A \cdot B = A \cdot C = 0$,

but $B \neq C$.
NOTE: This says that the cancellation property does not hold for matrix multiplication.
Find three non-zero matrices E, F, G, different than those given above such that $E \cdot F = E \cdot G$, but $F \neq G$.

---

1. *To express a system of linear equations in matrix form and to introduce the coefficient matrix and augmented matrix of the system.*
2. *To solve a system of linear equations in matrix form.*
3. *To discuss solutions of a system of linear equations in terms of its coefficent matrix.*

Consider the following linear system:

$$
\text{(I)}\quad
\begin{array}{c}
a_{11}x_1 + a_{12}x_2 + \ldots + a_{1n}x_n = b_1 \\
a_{21}x_1 + a_{22}x_2 + \ldots + a_{2n}x_n = b_2 \\
\vdots \qquad\qquad \vdots \qquad\qquad \vdots \qquad \vdots \\
a_{m1}x_1 + a_{m2}x_2 + \ldots + a_{mn}x_n = b_m
\end{array}
$$

The matrix
$$
A = \begin{bmatrix} a_{11} & a_{12} & \ldots & a_{1n} \\ \vdots & \vdots & & \vdots \\ a_{m1} & a_{mn} & \ldots & a_{mn} \end{bmatrix}
$$
whose entries are the coefficients of the system is called the coefficient matrix of the system and the matrix

$$
\left[\begin{array}{cccc:c} a_{11} & a_{12} & \ldots & a_{1n} & b_1 \\ \vdots & \vdots & \vdots & & \vdots \\ a_{m1} & a_{m2} & \ldots & a_{mn} & b_m \end{array}\right]
$$

is called the augmented matrix of the system. Thus a system with *m* equations and *n* unknowns has a *m×n* coefficient matrix and a *m×(n+1)* augmented matrix.

For example, the system

$$3x + 2y = 4$$
$$5x - y = -3$$

has coefficient natrix $\begin{bmatrix} 3 & 2 \\ 5 & -1 \end{bmatrix}$ and augmented matrix $\begin{bmatrix} 3 & 2 & \vdots & 4 \\ 5 & -1 & \vdots & -3 \end{bmatrix}$.

The system

$$-3x + 2y + 5z = 4$$
$$4x - 5y + z = 1$$

has coefficent matrix $\begin{bmatrix} -3 & 2 & 5 \\ 4 & -5 & 1 \end{bmatrix}$ and augmented matrix

$\begin{bmatrix} -3 & 2 & 5 & \vdots & 4 \\ 4 & -5 & 1 & \vdots & 1 \end{bmatrix}$.

The linear system (I) can be expressed in matrix form as follows:

$$AX = B,$$

where A is the coeeficient matrix,

$$X = \begin{bmatrix} x_1 \\ x_2 \\ \vdots \\ x_n \end{bmatrix}, \quad B = \begin{bmatrix} b_1 \\ \vdots \\ b_m \end{bmatrix}$$

Thus,

$$AX = \begin{bmatrix} a_{11} & a_{12} \cdots & a_{1n} \\ \vdots & \vdots & \vdots \\ a_{m1} & a_{m2} \cdots & a_{mn} \end{bmatrix} \begin{bmatrix} x_1 \\ \vdots \\ x_n \end{bmatrix}$$

$$= \begin{bmatrix} a_{11}x_1 + a_{12}x_2 + \ldots + a_{1n}x_n \\ \vdots \qquad \vdots \qquad \vdots \\ a_{m1}x_1 + a_{m2}x_2 + \ldots + a_{mn}x_n \end{bmatrix} = \begin{bmatrix} b_1 \\ \vdots \\ b_m \end{bmatrix} = B$$

111

As an example, the system

$$3x + 2y = 4$$
$$5x - y = -3$$

has the matrix form $\begin{bmatrix} 3 & 2 \\ 5 & -1 \end{bmatrix} \begin{bmatrix} x \\ y \end{bmatrix} = \begin{bmatrix} 4 \\ -3 \end{bmatrix}$

SELF-TEST:

(a) Change the following system to matrix form, and find its augmented and coefficient matrices:

$$-2x_1 + 3x_2 + x_3 = -5$$
$$x_1 - x_2 + 3x_3 = 0$$
$$4x_1 + 2x_2 - 5x_3 = -2$$

(b) If the augmented matrix of a system is

$$\begin{bmatrix} 0 & -1 & 2 & | & 2 \\ 1 & 3 & -1 & | & 1 \\ 5 & 7 & -3 & | & 3 \\ 2 & -1 & 0 & | & 4 \end{bmatrix}$$

Write the system in equation form.

ANS: (a) The matrix form $\begin{bmatrix} -2 & 3 & 1 \\ 1 & -1 & 3 \\ 4 & 2 & -5 \end{bmatrix} \begin{bmatrix} x_1 \\ x_2 \\ x_3 \end{bmatrix} = \begin{bmatrix} -5 \\ 0 \\ -2 \end{bmatrix}$

Augmented matrix: $\begin{bmatrix} -2 & 3 & 1 & | & -5 \\ 1 & -1 & 3 & | & 0 \\ 4 & 2 & -5 & | & -2 \end{bmatrix}$

Coefficient matrix: $\begin{bmatrix} -2 & 3 & 1 \\ 1 & -1 & 3 \\ 4 & 2 & -5 \end{bmatrix}$

(b) $- y + 2z = 2$
$x + 3y - z = 1$
$5x + 7y - 3z = 3$
$2x - y = 4$

A system of linear equations can be uniquely represented in terms of its augmented matrix. To solve a system, we only have to manipulate the augmented matrix as will be explained below.

In Chapter 1 we discussed the method of elimination to solve a system of linear equations. In this section we show that the method of elimination is equivalent to changing the augmented matrix to row-reduced form. Recall that the operations allowed to change a system into its equivalence are:

(i)   interchanging two equations,
(ii) multiplying an equation by a non-zero constant,
(iii) adding a multiple of one equation to another,

If each equation of the system is considered as a row of its augmented matrix, then the three above mentioned operations are precisely the same as the elementary row operations.

As an example, consider the system

$x + 3y = -2$
$2x - y = 3$

The augmented matrix of the system is

$$\begin{bmatrix} 1 & 3 & \vdots & -2 \\ 2 & -1 & \vdots & 3 \end{bmatrix}$$

The row-reduced form is

113

$$\begin{bmatrix} 1 & 0 & \vdots & 1 \\ 0 & 1 & \vdots & -1 \end{bmatrix}$$

which is the augmented matrix of a system equivalent to the given system. The solution of the second system is x = 1 and y = -1. Therefore, the solution of the original system is x = 1 and y = -1.

EXAMPLE 1. Solve the system in matrix form

$$2x - 3y + 2z = 9$$
$$x + 2y - z = -3$$
$$3x - y + 5z = 14$$

Solution: The augmented matrix is

$$\begin{bmatrix} 2 & -3 & 2 & \vdots & 9 \\ 1 & 2 & -1 & \vdots & -3 \\ 3 & -1 & 5 & \vdots & 14 \end{bmatrix}$$

Apply elementary row operations repeatedly we have

$$\begin{bmatrix} 2 & -3 & 2 & \vdots & 9 \\ 1 & 2 & -1 & \vdots & -3 \\ 3 & -1 & 5 & \vdots & 14 \end{bmatrix} \rightarrow \begin{bmatrix} 1 & 2 & -1 & \vdots & -3 \\ 2 & -3 & 2 & \vdots & 9 \\ 3 & -1 & 5 & \vdots & 14 \end{bmatrix}$$

$$\rightarrow , \begin{bmatrix} 1 & 2 & -1 & \vdots & -3 \\ 0 & -7 & 4 & \vdots & 15 \\ 0 & -7 & 8 & \vdots & 23 \end{bmatrix} \rightarrow \begin{bmatrix} 1 & 2 & -1 & \vdots & -3 \\ 0 & -7 & 4 & \vdots & 15 \\ 0 & 0 & 4 & \vdots & 8 \end{bmatrix}$$

$$\rightarrow \begin{bmatrix} 1 & 2 & -1 & \vdots & -3 \\ 0 & 1 & -\frac{4}{7} & \vdots & \frac{-15}{7} \\ 0 & 0 & 1 & \vdots & 2 \end{bmatrix} \rightarrow \begin{bmatrix} 1 & 0 & 0 & \vdots & 1 \\ 0 & 1 & 0 & \vdots & -1 \\ 0 & 0 & 1 & \vdots & 2 \end{bmatrix}$$

Therefore, the solution of the system is x = 1, y = -1, z = 2.

EXAMPLE 2. Solve the system in matrix form

$$x_1 + 2x_2 - x_3 + 4x_4 = 11$$
$$2x_1 + 5x_2 + x_3 + 6x_4 = 13$$
$$3x_1 + 7x_2 + 2x_3 + 14x_4 = 22$$

The system in matrix form is

$$\begin{bmatrix} 1 & 2 & -1 & 4 & \vdots & 11 \\ 2 & 5 & 1 & 6 & \vdots & 13 \\ 3 & 7 & 2 & 14 & \vdots & 22 \end{bmatrix}$$

which is row equivalent to the following row-reduced form

$$\begin{bmatrix} 1 & 0 & 0 & 22 & \vdots & 22 \\ 0 & 1 & 0 & -8 & \vdots & -6 \\ 0 & 0 & 1 & 2 & \vdots & -1 \end{bmatrix}$$

and the corresponding solutions are:

$$x_4 = \text{arbitrary}$$
$$x_1 = 22 - 22x_4$$
$$x_2 = -6 + 8x_4$$
$$x_3 = -1 - 2x_4$$

EXAMPLE 3. Solve the system in matrix form

$$x + 2y + z = 3$$
$$2x - y + 3z = 2$$
$$3x + y + 4z = 1$$

Solution: The augmented matrix is

$$\begin{bmatrix} 1 & 2 & 1 & \vdots & 3 \\ 2 & -1 & 3 & \vdots & 2 \\ 3 & 1 & 4 & \vdots & 1 \end{bmatrix}$$

Change it to row-equivalent form to obtain

115

$$\begin{bmatrix} 1 & 2 & 1 & \vdots & 3 \\ 0 & -5 & 1 & \vdots & -4 \\ 0 & -5 & 1 & \vdots & -8 \end{bmatrix} \rightarrow \begin{bmatrix} 1 & 2 & 1 & \vdots & 3 \\ 0 & -5 & 1 & \vdots & -4 \\ 0 & 0 & 0 & \vdots & -4 \end{bmatrix}$$

The last row corresponds to the equation

$$0 = -4$$

Thus the system has no solution.

If the system is homogeneous, i.e., the constants of
of each equations are zero, to solve the system in matrix form,
we only have to deal with its coefficient matrix. For example,
consider the homogenous system

$$\begin{aligned} x + 2y - z &= 0 \\ 2x - y + 5z &= 0 \\ 3x + 4y + 6z &= 0 \end{aligned}$$

The coefficient matrix is

$$\begin{bmatrix} 1 & 2 & -1 \\ 2 & -1 & 5 \\ 3 & 4 & 6 \end{bmatrix}$$

It is row-equivalent to the row-reduced form

$$\begin{bmatrix} 1 & 0 & 0 \\ 0 & 1 & 0 \\ 0 & 0 & 1 \end{bmatrix}$$

Thus, the system has only the trivial solution.

EXAMPLE 4. Solve the system of homogeous linear equations

$$\begin{aligned} x + y - z &= 0 \\ 2x - 2y + 3z &= 0 \\ 3x - y + 2z &= 0 \end{aligned}$$

116

Solution: The coefficient matrix is:

$$\begin{bmatrix} 1 & 1 & -1 \\ 2 & -2 & 3 \\ 3 & -1 & 2 \end{bmatrix}$$

which is equivalent to

$$\begin{bmatrix} 1 & 1 & -1 \\ 0 & 1 & -\frac{5}{4} \\ 0 & 0 & 0 \end{bmatrix}$$

Thus, the solutions are

$$z = \text{arbitrary}$$
$$y = \frac{5}{4}z$$
$$x = -\frac{1}{4}z$$

SELF-TEST:

Solve each of the following systems in matrix form:

(a)
$$x - y + 2z = -1$$
$$2x + 2y - z = 0$$
$$x + y = 5$$
$$2x + 2z = 4$$

(b)
$$u + v + w = 3$$
$$2u - 2v + 2w = 2$$

(c)
$$x + y - z = 2$$
$$2x - y + 3z = 4$$
$$4x + y + z = 3$$

(d)
$$2x - y + z - w = 0$$
$$x + 2y - z + 2w = 0$$
$$5x - y + 6z - 3w = 0$$
$$-2x + 3y + 2z - w = 0$$

ANS:

(a) $\begin{bmatrix} x \\ y \\ z \end{bmatrix} = \begin{bmatrix} -8 \\ 13 \\ 10 \end{bmatrix}$

117

(b) The system is equivalent to

$$\left[\begin{array}{ccc|c} 1 & 0 & 1 & 2 \\ 0 & 1 & 0 & 1 \end{array}\right]$$

then the solutions are:

w = arbitrary

$v = 1$

$u = 2 - w$

(c) no solution

(d) The system has only the trivial solution since it is equivalent to the identity matrix.

If a system has the same number of equations as unknowns, the number of solutions can be related to its coefficient matrix. In fact, the following theorem is true.

THEOREM 1. If the square matrix Z is the coefficient matrix of a linear system, then the system has a unique solution if and only if A is invertible.

If $AX = B$ is the matrix form of a system and A is invertible then the solution of the system is

$$X = A^{-1}B$$

EXAMPLE 5. The system

$$\left[\begin{array}{cc} 2 & 1 \\ 1 & 1 \end{array}\right] \left[\begin{array}{c} x \\ y \end{array}\right] = \left[\begin{array}{c} 1 \\ 3 \end{array}\right]$$

has the unique solution

$$\begin{bmatrix} x \\ y \end{bmatrix} = \begin{bmatrix} 2 & 1 \\ 1 & 1 \end{bmatrix}^{-1} \begin{bmatrix} 1 \\ 3 \end{bmatrix}$$

$$= \begin{bmatrix} 1 & -1 \\ -1 & 2 \end{bmatrix} \begin{bmatrix} 1 \\ 3 \end{bmatrix} = \begin{bmatrix} -2 \\ 5 \end{bmatrix}$$

EXAMPLE 6. The system

$$\begin{bmatrix} 1 & 3 & 3 \\ 1 & 4 & 3 \\ 1 & 3 & 4 \end{bmatrix} \begin{bmatrix} x \\ y \\ z \end{bmatrix} = \begin{bmatrix} -1 \\ 2 \\ 3 \end{bmatrix}$$

has the unique solution

$$\begin{bmatrix} x \\ y \\ z \end{bmatrix} = \begin{bmatrix} 1 & 3 & 3 \\ 1 & 4 & 3 \\ 1 & 3 & 4 \end{bmatrix}^{-1} \begin{bmatrix} -1 \\ 2 \\ 3 \end{bmatrix}$$

$$= \begin{bmatrix} 7 & -3 & -3 \\ -1 & 1 & 0 \\ -1 & 0 & 1 \end{bmatrix} \begin{bmatrix} -1 \\ 2 \\ 3 \end{bmatrix}$$

$$= \begin{bmatrix} -22 \\ 3 \\ 4 \end{bmatrix}$$

In case the coefficient matrix is not invertible the system may have no solution or infinitely many solutions as can be seen from the following examples:

EXAMPLE 7. Consider the system

$$\begin{bmatrix} 1 & 2 \\ 2 & 4 \end{bmatrix} \begin{bmatrix} x \\ y \end{bmatrix} = \begin{bmatrix} 3 \\ 6 \end{bmatrix}$$

The coefficient matrix $\begin{bmatrix} 1 & 2 \\ 2 & 4 \end{bmatrix}$ is not invertible but the system

119

has infinitely many solutions:

$$y = \text{arbitrary}$$
$$x = 3 - 2y$$

EXAMPLE 8. The system

$$\begin{bmatrix} 2 & -3 \\ -4 & 6 \end{bmatrix} \begin{bmatrix} x \\ y \end{bmatrix} = \begin{bmatrix} 1 \\ 3 \end{bmatrix}$$

with non-invertible coefficient matrix has no solution.

In case the system is homogeneous, we have the following theorem:

THEOREM 2. If a homogeneous system has the same number of equations and unknowns, then

1. the system has only the trivial solution if the coefficient matrix is invertible.
2. the system has infinitely many solutions if the coefficient matrix is not invertible.

EXAMPLE 9. The homogeneous system

$$\begin{bmatrix} 1 & 2 & -1 \\ 1 & 3 & 2 \\ 2 & 5 & 1 \end{bmatrix} \begin{bmatrix} x \\ y \\ z \end{bmatrix} = \begin{bmatrix} 0 \\ 0 \\ 0 \end{bmatrix}$$

has infintely many solutions because the coefficient matrix

$$\begin{bmatrix} 1 & 2 & -1 \\ 1 & 3 & 2 \\ 2 & 5 & 1 \end{bmatrix}$$

can be reduced to

$$\begin{bmatrix} 1 & 0 & -7 \\ 0 & 1 & 3 \\ 0 & 0 & 0 \end{bmatrix}$$

and is thus not invertible.

The following theorem is very important for systems of homogeneous linear equations.

THEOREM 3.   If the number of unknowns is more than the number of equations in a homogeneous linear system, then the system has infintely many solutions.

For example, the system

$$2x_1 + 3x_2 - 4x_3 + x_4 = 0$$
$$x_1 - 2x_2 - 3x_3 + 2x_4 = 0$$
$$2x_1 - 5x_2 + 2x_3 - x_4 = 0$$

has infinitely many solutions since the number of unknowns is 4, which is greater than 3, the number of equations in the system.

SELF-TEST:

1. Based on the coefficient matrix of each of the following systems, determine if each system has a unique solution or otherwise.

(a) $2x - y = 3$
       $3x + 5y = 7$

(b) $-3x + 2y = 5$
       $6x - 4y = 7$

(c) $\begin{bmatrix} 1 & 4 & -3 \\ 2 & 0 & 1 \\ -1 & 3 & 2 \end{bmatrix} \begin{bmatrix} x \\ y \\ z \end{bmatrix} = \begin{bmatrix} -2 \\ 0 \\ 1 \end{bmatrix}$

(d) $\begin{bmatrix} -2 & 1 & 0 \\ 1 & 3 & 7 \\ -1 & 4 & 7 \end{bmatrix} \begin{bmatrix} u \\ v \\ w \end{bmatrix} = \begin{bmatrix} -1 \\ 1 \\ 3 \end{bmatrix}$

2. For each of the following homogeneous systems, determine
    if it has a unique solution or infintely many solutions.

(a) $\begin{bmatrix} 1 & -1 & 2 \\ 3 & 2 & 0 \\ -2 & 1 & 1 \end{bmatrix} \begin{bmatrix} x \\ y \\ z \end{bmatrix} = \begin{bmatrix} 0 \\ 0 \\ 0 \end{bmatrix}$ (b) $\begin{bmatrix} 3 & -1 & 5 \\ 2 & 0 & 1 \\ -1 & -1 & 3 \end{bmatrix} \begin{bmatrix} x \\ y \\ z \end{bmatrix} = \begin{bmatrix} 0 \\ 0 \\ 0 \end{bmatrix}$

(c) $\begin{bmatrix} 1 & 2 & 4 & 3 \\ -1 & 0 & 2 & 3 \\ 2 & 1 & 0 & 4 \end{bmatrix} \begin{bmatrix} x \\ y \\ z \\ w \end{bmatrix} = \begin{bmatrix} 0 \\ 0 \\ 0 \end{bmatrix}$

EXERCISES 2.4

In Exercises 1-3, find the coefficient and augmented matrix for
the following system. Also, express the system in matrix form.

1. $2x_1 + 3x_2 - 5x_3 = -2$
    $x_1 - 4x_2 + 6x_3 = -1$
    $2x_1 + 2x_2 - 6x_3 = 5$

2. $2x + 3y = -1$
    $4x - y = 5$

3. $6x_1 - 3x_2 + 4x_3 - 2x_4 = -1$
    $2x_1 + 2x_2 - x_3 + 5x_4 = 2$
    $-12x_1 + x_3 - 5x_4 = 3$

4. If the augmented matrix of a system is

$$\begin{bmatrix} 2 & -1 & \vdots & 2 \\ 3 & 4 & \vdots & 5 \end{bmatrix}$$

write the system in equation form and in matrix form.

5. Same question as in Exercise 4 with augmented matrix

$$\begin{bmatrix} -1 & 2 & 0 & 3 & \vdots & 0.1 \\ 15 & 7 & 4 & 9 & \vdots & -0.5 \\ -5 & 6 & 0.8 & 7 & \vdots & -1.9 \end{bmatrix}$$

In Exercises 6-10, solve each of the systems, which are expressed as an augmented matrix:

6. $$\begin{bmatrix} 2 & 3 & \vdots & 0 \\ -1 & 2 & \vdots & -1 \end{bmatrix}$$

7. $$\begin{bmatrix} 1 & -1 & 5 & \vdots & 1 \\ 0 & 2 & 1 & \vdots & 2 \\ -1 & 3 & 1 & \vdots & 3 \end{bmatrix}$$

8. $$\begin{bmatrix} 1 & 1 & -1 & 2 & \vdots & 1 \\ 0 & -1 & 1 & 0 & \vdots & 1 \\ -1 & 1 & -1 & 1 & \vdots & 2 \end{bmatrix}$$

9. $$\begin{bmatrix} 1 & -1 & 3 & 0 & \vdots & -1 \\ -1 & 2 & 0 & 4 & \vdots & 1 \\ 0 & 1 & 3 & 4 & \vdots & 3 \end{bmatrix}$$

10. $$\begin{bmatrix} 2 & -1 & 1 & 3 & \vdots & 0 \\ 1 & 0 & 2 & -1 & \vdots & 0 \\ 1 & -1 & 1 & 3 & \vdots & 0 \\ -1 & 1 & 2 & 1 & \vdots & 0 \end{bmatrix}$$

11. Based on the coefficient matrix of the following system

determine if it has a unique solution.

$$x + 2y - z = 2$$
$$-x + y + z = 1$$
$$2x - y - z = -1$$

12. Same question as in Exercise 11 for the system

$$2x - y + z + w = -1$$
$$x + 2y + z - 3w = -2$$
$$3x - y + z + 4w = 1$$
$$5x - 2y + 2z + 5w = 0$$

13. Same question as in Exercise 11 for the system

$$\begin{bmatrix} 2 & -1 & 1 & 1 & \vdots & 0 \\ 1 & 0 & 1 & 1 & \vdots & 0 \\ -2 & 0 & 1 & -1 & \vdots & 0 \\ 1 & 1 & -1 & 1 & \vdots & 0 \end{bmatrix}$$

In Exercise 14-16, based on the coefficient matrix of each homogeneous system, determine if the system has a unique solution or infinitely many solutions.

14. $$\begin{bmatrix} 1 & 1 & 2 & -1 & \vdots & 0 \\ 3 & 1 & 0 & 1 & \vdots & 0 \\ 9 & 15 & 24 & -28 & \vdots & 0 \end{bmatrix}$$

15. $$\begin{bmatrix} 1 & 1 & 1 & \vdots & 0 \\ 2 & 2 & 3 & \vdots & 0 \\ 3 & 3 & 5 & \vdots & 0 \end{bmatrix}$$

16. $$\begin{bmatrix} 1 & -2 & 1 & \vdots & 0 \\ 3 & -1 & 0 & \vdots & 0 \\ 2 & 1 & -1 & \vdots & 0 \\ 4 & -3 & 1 & \vdots & 0 \end{bmatrix}$$

2.5 SOME APPLICATIONS

OBJECTIVES:

1. To discuss some applications of matrices

EXAMPLE 1. A fast food chain has three restaurants located in cities A, B, C. They all sell the same four different foods, I, II, III, IV. During the month of July, their sales amounts are given by the following matrix

	I	II	III	IV
A	50	40	35	60
B	60	35	70	40
C	45	65	20	25

The price and cost measured in $100 units for each food is given by the matrix

	PRICE	COST
I	2	1
II	3	2
III	4	2
IV	3	1

What is gross income and profit for each of the restaurants during the month of July? The answer can be easily found by multiplying two matrices

$$\begin{bmatrix} 50 & 40 & 35 & 60 \\ 60 & 35 & 70 & 40 \\ 45 & 65 & 20 & 25 \end{bmatrix} \begin{bmatrix} 2 & 1 \\ 3 & 2 \\ 4 & 2 \\ 3 & 1 \end{bmatrix}$$

$$= \begin{bmatrix} 540 & 260 \\ 625 & 310 \\ 440 & 240 \end{bmatrix} \begin{matrix} A \\ B \\ C \end{matrix}$$

GROSS   TOTAL
INCOME   COST

The profit for each restaurant is

$$\begin{matrix} A \\ B \\ C \end{matrix} \begin{bmatrix} 540-260 = 280 \\ 625-310 = 315 \\ 440\ 240 = 200 \end{bmatrix}$$

EXAMPLE 2. Leontief input-output analysis was developed by Wassily Leontief who received the Nobel prize in economics in 1973. It's purpose is to describe a simple economy in which consumption is equal to production.

Assume an economy has n basic commodities. The gross production of each commodity is represented by the column matri:

$$X = \begin{bmatrix} x_1 \\ x_2 \\ \vdots \\ x_n \end{bmatrix}$$

and each entry of the matrix

$$A = [a_{ij}]_{n \times n}$$

is the amount of commodity i consumed to produce one unit of commodity j. We shall call the matrix A the input-output coefficient matrix. Let the column matrix

$$Y = \begin{bmatrix} y_1 \\ y_2 \\ \vdots \\ y_n \end{bmatrix}$$

represent the final demand (or net production) of each commodity. Since the gross product for each commodity must be equal to the sum of the total amounts used by producing other commodities and the net production, we thus have the relation

$$x_1 = a_{11}x_1 + a_{12}x_2 + .. + a_{1n}x_n + y_1$$
$$x_2 = a_{21}x_1 + a_{22}x_2 + .. + a_{2n}x_n + y_2$$
$$\vdots \qquad \vdots \qquad \vdots \qquad \qquad \vdots \qquad \vdots$$
$$x_n = a_{n1}x_1 + a_{n2}x_2 + .. + a_{nn}x_n + y_n$$

This can be expressed in matrix notation as follows:

$$\begin{bmatrix} x_1 \\ \vdots \\ x_n \end{bmatrix} = \begin{bmatrix} a_{11} & \cdots & a_{1n} \\ \vdots & & \vdots \\ a_{n1} & \cdots & a_{nn} \end{bmatrix} \begin{bmatrix} x_1 \\ \vdots \\ x_n \end{bmatrix} + \begin{bmatrix} y_1 \\ \vdots \\ y_n \end{bmatrix}$$

or, $X = AX + Y$.

The above equation can be rewritten as

$$I_n X - AX = Y \quad \text{or} \quad (I_n - A)X = Y$$

If the matrix $I_n - A$ is invertible, multiplying both sides of the above expression by $(I_n - A)^{-1}$, we have

$$X = (I_n - A)^{-1} Y$$

Therefore, if we know the input-ouput coefficient matrix A, and the amounts of final demand for all commodities Y_i, we can find the gross production X_i for each commodity. As an example, assume the input-output coefficient matrix for commodities I, II, III is given by

$$A = \begin{bmatrix} 0 & \frac{1}{2} & \frac{1}{3} \\ \frac{1}{2} & 0 & \frac{1}{4} \\ \frac{1}{3} & \frac{1}{2} & 0 \end{bmatrix}$$

and the final demand is given by

$$Y = \begin{bmatrix} 1680 \\ 2240 \\ 1960 \end{bmatrix}$$

Then

$$I_n - A = \begin{bmatrix} 1 & -\frac{1}{2} & -\frac{1}{3} \\ -\frac{1}{2} & 1 & -\frac{1}{4} \\ -\frac{1}{3} & -\frac{1}{2} & 1 \end{bmatrix}$$

$$(I_n - A)^{-1} = \frac{1}{28} \begin{bmatrix} 63 & 48 & 33 \\ 42 & 64 & 30 \\ 42 & 48 & 54 \end{bmatrix}$$

$$(I_n - A)^{-1} Y = \frac{1}{28} \begin{bmatrix} 63 & 48 & 33 \\ 42 & 64 & 30 \\ 42 & 48 & 54 \end{bmatrix} \begin{bmatrix} 1680 \\ 2240 \\ 1960 \end{bmatrix}$$

$$= \begin{bmatrix} 9930 \\ 9740 \\ 10140 \end{bmatrix}$$

which says the gross production for commodity I should be 9,930 units, for commodity II 9,740 units and for commodity III 10,140 units.

EXAMPLE 3. SECRET CODES: Messages can be sent via some secret codes. There are different ways for designing codes to preserve secrecy. Here we introduce a method of creating codes based on multiplication of matrices with integer entries. We first associate the natural numbers 1, 2, ..., 26 with letters of the alphabet in the following way:

```
A   B   C ................Z
↕   ↕   ↕                  ↕
1   2   3                  26
```

To code a message, we select an invertible matrix A such that both A and its inverse have integral entries. As an example,

$$\text{let } A = \begin{bmatrix} 3 & 1 \\ 2 & 1 \end{bmatrix}, \text{ then } A^{-1} = \begin{bmatrix} 1 & -1 \\ -2 & 3 \end{bmatrix}$$

Suppose the message to be sent is the following:

TOMORROW IS THE DAY

The corresponding numbers are 20, 15, 13, 15, 18, 18, 15, 23, 9, 19, 20, 8, 5, 4, 1, 25. Write each pair of numbers as a column matrix and multiply each of the column matrices from left by the matrix A. We have

$$\begin{bmatrix} 3 & 1 \\ 2 & 1 \end{bmatrix} \begin{bmatrix} 20 \\ 15 \end{bmatrix} = \begin{bmatrix} 75 \\ 55 \end{bmatrix}$$

$$\begin{bmatrix} 3 & 1 \\ 2 & 1 \end{bmatrix} \begin{bmatrix} 13 \\ 15 \end{bmatrix} = \begin{bmatrix} 54 \\ 41 \end{bmatrix}$$

$$\begin{bmatrix} 3 & 1 \\ 2 & 1 \end{bmatrix} \begin{bmatrix} 18 \\ 18 \end{bmatrix} = \begin{bmatrix} 72 \\ 54 \end{bmatrix}$$

$$\begin{bmatrix} 3 & 1 \\ 2 & 1 \end{bmatrix} \begin{bmatrix} 15 \\ 23 \end{bmatrix} = \begin{bmatrix} 68 \\ 53 \end{bmatrix}$$

$$\begin{bmatrix} 3 & 1 \\ 2 & 1 \end{bmatrix} \begin{bmatrix} 9 \\ 19 \end{bmatrix} = \begin{bmatrix} 46 \\ 37 \end{bmatrix}$$

$$\begin{bmatrix} 3 & 1 \\ 2 & 1 \end{bmatrix} \begin{bmatrix} 20 \\ 8 \end{bmatrix} = \begin{bmatrix} 68 \\ 48 \end{bmatrix}$$

$$\begin{bmatrix} 3 & 1 \\ 2 & 1 \end{bmatrix} \begin{bmatrix} 5 \\ 4 \end{bmatrix} = \begin{bmatrix} 19 \\ 14 \end{bmatrix}$$

$$\begin{bmatrix} 3 & 1 \\ 2 & 1 \end{bmatrix} \begin{bmatrix} 1 \\ 25 \end{bmatrix} = \begin{bmatrix} 28 \\ 27 \end{bmatrix}$$

The coded message is:

75   55   54   41   72   54   68   53   46   37   68
48   19   14   28   27

To decode the message we multiply each column matrix from the left by A^{-1}:

$$A^{-1}\begin{bmatrix} 75 \\ 55 \end{bmatrix} = \begin{bmatrix} 1 & -1 \\ -2 & 3 \end{bmatrix} \begin{bmatrix} 75 \\ 55 \end{bmatrix} = \begin{bmatrix} 20 \\ 15 \end{bmatrix}$$

$$A^{-1}\begin{bmatrix} 54 \\ 41 \end{bmatrix} = \begin{bmatrix} 1 & -1 \\ -2 & 3 \end{bmatrix} \begin{bmatrix} 54 \\ 41 \end{bmatrix} = \begin{bmatrix} 13 \\ 15 \end{bmatrix}$$

Thus the original message can be received.

Of course it is not necessary to set the association between letters and numbers as $A \leftrightarrow 1$, $B \leftrightarrow 2$, etc. One may choose any 1 - 1 correspondence.

Also, we can select 3x3, 4x4 or higher order matrices. But it is important that the matrix chosen be invertible and the matrix and its inverse have integral entries. For example, if we want to send the same message with the same 1 - 1 correspondence between letter and numbers but use the 3x3 matrix

$$B = \begin{bmatrix} 1 & 3 & 3 \\ 1 & 4 & 3 \\ 1 & 3 & 4 \end{bmatrix}$$

Instead of writing a pair of numbers we write three numbers as a column matrix. Since there are 16 numbers in total, we add 2 numbers 26 and 26 (Z) to make it a multiple of 3. The encoded message can be obtained as

$$\begin{bmatrix} 1 & 3 & 3 \\ 1 & 4 & 3 \\ 1 & 3 & 4 \end{bmatrix} \begin{bmatrix} 20 \\ 15 \\ 13 \end{bmatrix} = \begin{bmatrix} 104 \\ 119 \\ 117 \end{bmatrix}, \quad \begin{bmatrix} 1 & 3 & 3 \\ 1 & 4 & 3 \\ 1 & 3 & 4 \end{bmatrix} \begin{bmatrix} 15 \\ 18 \\ 18 \end{bmatrix} = \begin{bmatrix} 123 \\ 141 \\ 141 \end{bmatrix}, \text{ etc.}$$

To decode the message, we use the inverse

$$B^{-1} = \begin{bmatrix} 7 & -3 & -3 \\ -1 & 1 & 0 \\ -1 & 0 & 1 \end{bmatrix}$$

EXERCISE 2.5

1. John and Mary are planning to buy a house before they get married. In order to come up with the 10% downpayment, they must sell some stocks they own. Suppose the number of shares of each stock they own are given by the 2x4 matrix

	STOCK 1	STOCK 2	STOCK 3	STOCK 4
John	50	20	100	25
Mary	30	40	50	50

The selling price for each share is given by

STOCK 1 $15.75
STOCK 2 $50.25
STOCK 3 $10.50
STOCK 4 $30.20

Express, in matrix form, the amounts of money received by John and Mary from selling all their stocks.

2. The ABC Company has three warehouses located in different cities. Each warehouse stores four different goods. In the month of January, the orders for each good from each warehouse are given as a 3x4 matrix:

$$\text{WAREHOUSE} \begin{bmatrix} 1500 & 500 & 200 & 1000 \\ 1000 & 400 & 500 & 800 \\ 1200 & 300 & 400 & 900 \end{bmatrix}$$

and the price for each good in given by

$$\begin{bmatrix} \$20 \\ \$50 \\ \$15 \\ \$10 \end{bmatrix}$$

Find in matrix form the total income for each warehouse. Also find the total income of the ABC Company during January.

3. Assume an economy has three commodities and the input-output coefficient matrix for the commodities is given by:

$$A = \begin{bmatrix} \frac{1}{5} & \frac{1}{3} & \frac{1}{4} \\ \frac{1}{10} & \frac{1}{7} & \frac{1}{5} \\ \frac{1}{3} & \frac{1}{2} & 0 \end{bmatrix}$$

and the final demand is

$$Y = \begin{bmatrix} 100 \\ 200 \\ 300 \end{bmatrix}$$

Find the gross production for each of the commodities.

4. Repeat Exercise 3 with the coefficient matrix:

$$A = \begin{bmatrix} 0 & \frac{1}{4} & \frac{1}{3} \\ \frac{1}{3} & 0 & \frac{1}{5} \\ \frac{1}{4} & \frac{1}{5} & 0 \end{bmatrix} \quad \text{and} \quad Y = \begin{bmatrix} 40 \\ 70 \\ 50 \end{bmatrix}$$

5. Use the correspondence

```
A   B   C .....  X   Y   Z
↕   ↕   ↕        ↕   ↕   ↕
1   2   3        24  25  26
```

and the matrices

$$A = \begin{bmatrix} 5 & 2 \\ 2 & 1 \end{bmatrix} \quad \text{and} \quad B = \begin{bmatrix} 1 & -1 & 1 \\ 2 & -3 & 1 \\ -1 & 1 & -2 \end{bmatrix}$$

to encode and then decode each of the following messages.

(a) TO BE OR NOT TO BE

(b) MATHEMATICS IS FUN

    (c)   INTEREST IS RISING

6.   Using the same correspondence given in Problem 5 and the matrix

$$A = \begin{bmatrix} 1 & 1 \\ 0 & 1 \end{bmatrix}$$

decode each of the following messages:

(a)   40   15   22   1   23   5   8   7   30   15   23   19   41
        21   9   5   34   20

(b)   29   15   32   9   39   20   13   4   29   9   18   5

1. Let $A = \begin{bmatrix} 1 & -1 & 3 \\ 5 & 4 & 0 \end{bmatrix}$.

   The order of A is _____. The (2,2)-entry of A is

   _____.

2. A general matrix written as $A = [a_{ij}]_{m \times n}$ means A is a

   matrix with _____ rows and _____ columns. The (i,j)-entry

   of A is _____.

3. A square matrix is a matrix with _____.

4. Let $A = [a_{ij}]_{n \times n}$ be a square matrix. The elements on the

   diagonal of A are _____.

5. Let $A = [a_{ij}]_{n \times n}$ be a symmetric matrix. Then the (i,j)-

   entry of A is equal to the _____ entry of A.

6. Give examples of an upper triangular matrix and a lower
   triangular matrix which are not diagonal.

7. Give an example of each of the following:

   (a) scalar matrix
   (b) column matrix (column vector)
   (c) row matrix (row vector)

8. If $A = [a_{ij}]_{m \times n}$, $B = [b_{ij}]_{r \times s}$ are two matrices, and A = B

   then what can you say about m, n, r, s, a_{ij} and b_{ij}.

9. Let $A = \begin{bmatrix} 3 & -1 & 2 \\ 0 & 3 & 5 \end{bmatrix}$, $B = \begin{bmatrix} 1 & -6 & 3 \\ 7 & 0 & 4 \end{bmatrix}$. Compute A + B.

10. If $A = \begin{bmatrix} 4 & 2 \\ -5 & 0 \end{bmatrix}$ and a = -3, compute a · A.

11. Let $A = \begin{bmatrix} 1 & 3 \\ 4 & 5 \\ -1 & 0 \end{bmatrix}$. Find the transpose A' of A.

12. A is symmetric if and only if _____.

13. Let $A = [a_1 \ a_2]$, $B = \begin{bmatrix} b_1 \\ b_2 \end{bmatrix}$. Compute A·B.

14. Let $A = [a_{ij}]_{m \times n}$, $B = [b_{ij}]_{n \times p}$ and $C = A B = [c_{ij}]_{n \times s}$. We have

    (a) $n = $ _____, $s = $ _____

    (b) $c_{ij} = $ _____

15. Find two matrices A, B such that A·B ≠ B·A.

16. Let $A = \begin{bmatrix} 1 & -1 \\ 2 & 3 \end{bmatrix}$, $B = \begin{bmatrix} 1 & 0 \\ 5 & 2 \end{bmatrix}$, $C = \begin{bmatrix} 3 & -1 \\ 0 & 4 \end{bmatrix}$.

    (a) Compute (A·B)·C and A·(B·C)

    (b) Compute (A+B)·C and A·B + B·C

17. Find a **non-zero** matrix A such that $A^2 = 0$.

18. Find **two non-zero** matrices A, B such that A·B = 0.

19. Find three non-zero matrices A, B, C such that B ≠ C and A·B = A·C.

20. List the three elementary row operations for matrices.

21. Let $A = \begin{bmatrix} 1 & 3 \\ -2 & 5 \\ 0 & 4 \end{bmatrix}$. Find the matrix B obtained from A by

adding -3 times the third row to the second row.

22. Determine which of the following matrices is in row-reduced form:

(a) $\begin{bmatrix} 1 & 0 & -1 \\ 0 & 0 & 1 \\ 0 & 0 & 0 \end{bmatrix}$ (b) $\begin{bmatrix} 0 & 0 & 1 \\ 0 & 0 & 0 \\ 0 & 0 & 0 \end{bmatrix}$

(c) $\begin{bmatrix} 0 & 1 & -1 & 0 \\ 0 & 0 & 0 & 1 \\ 0 & 0 & 0 & 0 \end{bmatrix}$ (d) $\begin{bmatrix} 0 & 1 & 0 \\ 1 & 0 & 0 \end{bmatrix}$

23. If A is a square matrix in row-reduced form and A has no zero rows, then A must be the _____ matrix.

24. Change each of the following matrices to **row-reduced form**.

(a) $\begin{bmatrix} 1 & 2 \\ -1 & 3 \end{bmatrix}$ (b) $\begin{bmatrix} 1 & -1 & 1 \\ 2 & 0 & 3 \\ -2 & 0 & 1 \end{bmatrix}$

(c) $\begin{bmatrix} 2 & -1 & 1 & 0 \\ 1 & -1 & 2 & 1 \\ 3 & -2 & 3 & 1 \end{bmatrix}$ (d) $\begin{bmatrix} 1 & 0 & 1 & 0 \\ -1 & 1 & -1 & 1 \\ 1 & 1 & -1 & -1 \\ 0 & 1 & 0 & 1 \end{bmatrix}$

25. A **matrix A is said to be invertible if there exists a matrix B such that** _____.

26. If A is invertible, then A must be a _____ matrix.

27. If A is invertible, then its inverse is _____ and is denoted by _____.

28. Let A = $\begin{bmatrix} a & b \\ c & d \end{bmatrix}$. Then A is invertible if and only if _____.

29. Let $A = \begin{bmatrix} a & b \\ c & d \end{bmatrix}$, the determinant of A is _____.

30. If $A = \begin{bmatrix} a & b \\ c & d \end{bmatrix}$, and the determinant of A $\neq$ 0, then

$A^{-1} = \begin{bmatrix} & \\ & \end{bmatrix}$.

31. Find the inverse of

(a) $\begin{bmatrix} 5 & 3 \\ 1 & 4 \end{bmatrix}$ (b) $\begin{bmatrix} 2 & 4 \\ -1 & -3 \end{bmatrix}$

32. Apply the row reduction algorithm to determine if each of the following matrices is invertible and find the inverse if it exists.

(a) $\begin{bmatrix} 2 & 1 & 0 \\ 1 & -1 & 1 \\ 3 & 1 & -1 \end{bmatrix}$ (b) $\begin{bmatrix} 1 & -1 & 1 & 0 \\ 0 & 1 & 2 & 1 \\ 2 & 1 & -1 & 1 \\ 1 & 2 & -2 & 1 \end{bmatrix}$

33. A system of linear equations

$$a_{11}x_1 + a_{12}x_2 + \ldots + a_{1n}x_n = b_1$$
$$a_{21}x_1 + a_{22}x_2 + \ldots + a_{2n}x_n = b_1$$
$$\vdots \qquad \vdots \qquad \qquad \vdots \qquad \vdots$$
$$a_{m1}x_1 + a_{m2}x_2 + \ldots + a_{mn}x_n = b_m$$

can be written in the matrix form:

The coefficient matrix of the system is:

and the augmented matrix is:

34. Solve the following system in matrix form:

138

$$x - s + 2t = -1$$
$$2x + 2s - t = 0$$
$$x + s = 5$$
$$2x + 2t = 4$$

35. Solve the system with augmented matrix

$$\begin{bmatrix} 1 & 3 & -1 & \vdots & 1 \\ 2 & -3 & 1 & \vdots & 2 \\ -1 & 1 & 2 & \vdots & 3 \end{bmatrix}$$

36. A system of linear equations with the same number of variables and equations has a unique solution if and only if

_____.

37. A system of homogeneous equations has infinitely many solutions if _____.

38. If the input-output coefficient matrix of an economy is given as

$$A = \begin{bmatrix} 0 & \frac{1}{3} & \frac{1}{2} \\ \frac{1}{3} & 0 & \frac{1}{4} \\ \frac{1}{2} & \frac{1}{3} & 0 \end{bmatrix}$$

and the final demand is given by

$$Y = \begin{bmatrix} 100 \\ 200 \\ 300 \end{bmatrix}$$

Find the column matrix representing the gross product.

39. Use the correspondence

139

```
A    B    C  . . . . . . . . . . . X    Y    Z

↕    ↕    ↕                        ↕    ↕    ↕

1    2    3                        24   25   26
```

and the matrix $A = \begin{bmatrix} 5 & 3 \\ 3 & 2 \end{bmatrix}$ to decode the message

GOOD LUCK.

1. Give an example of a matrix which is upper triangular and symmetric.

2. Given $A = \begin{bmatrix} 2 & -1 & 3 \\ 4 & 2 & 0 \end{bmatrix}$, $B = \begin{bmatrix} 1 & 0 \\ -1 & 4 \\ 2 & 3 \end{bmatrix}$, $C = \begin{bmatrix} -1 & 1 \\ 2 & 3 \end{bmatrix}$

   compute each of the following that exists:

   (a) $A \cdot B + C$ (b) $(B \cdot C)'$

   (c) $A \cdot C$ (d) $C \cdot A$

3. Change the following matrix into row-reduced form

   $$\begin{bmatrix} -1 & 1 & 0 & 2 \\ 1 & -1 & 1 & 3 \\ 0 & 2 & -1 & 1 \end{bmatrix}$$

4. Determine all values of x so that the following matrix is not invertible

   $$\begin{bmatrix} x & x+2 \\ 1 & x \end{bmatrix}$$

5. Determine if the following matrix is invertible and find its inverse if it exists

   $$\begin{bmatrix} 2 & -1 & 0 \\ 1 & 1 & -1 \\ 0 & -1 & 2 \end{bmatrix}$$

6 Find all the 2x2 matrices in row-reduced form.

.7. Let $A = \begin{bmatrix} 1 & -1 \\ 2 & 0 \\ 3 & -2 \end{bmatrix}$, $B = \begin{bmatrix} 1 & -1 \\ 1 & -2 \\ 2 & 0 \end{bmatrix}$ How is B obtained from A?

8. Solve the system in matrix form

$$x + 2y - z + 4w = 11$$
$$2x + 5y + z + 6w = 13$$
$$3x + 7y + 2z + 14w = 32$$

9. Explain why a square matrix with a zero row or a zero column cannot be invertible.

10. (a) Find two matrices A and B such that $A \cdot B = 0$ and $B \cdot A \neq 0$

   (b) Find two matrices A and B such that $A \cdot B$ is an identity matrix but $B \cdot A$ is not.

   HINT: Consider non-square matrices.

COMPUTER APPLICATION: CHAPTER TWO

1. Write a program to read in ten 3x3 matrices and determine if each is

   (a) upper or lower triangular
   (b) diagonal
   (c) scalar
   (d) identity
   (e) symmetric

2. Write a program to perform matrix transpositions.

3. Write a program to perform the three elementary row operations.

4. Write a program to read in five 2x2 matrices and determine if they are invertible, Find the inverses that exist.

5. Write a program to change a 3x3 matrix to row-reduced form.

6. Write a program to solve a system of 3x3 linear equations by row-reduction method of matrices.

7. Write a program to apply the row-reduction method to determine if a 3x3 matrix is invertible.

8. A square matrix with positive integer entries is called a magic square if the sum of each row and the sum of each column are the same. As an example, the 3x3 matrix

$$\begin{bmatrix} 1 & 2 & 3 \\ 2 & 3 & 1 \\ 3 & 1 & 2 \end{bmatrix}$$

   is a magic square since the row sums and the column sums are equal. Write a program to find all 3x3 magic squares.

CHAPTER THREE

LINEAR PROGRAMMING

3.1 INTRODUCTION

OBJECTIVES:

1. *Introduce the objective function and the constraints of a linear programming problem.*
2. *State the general linear programming problem and introduce the concept of feasible solutions.*

Linear programming was developed during world war II to solve some resource allocation problem for the U.S. Air Force. It has become one of the most important tools to deal with problems in today's business and scientific world which involve either **maximizing** or **minimizing** a linear function subject to certain **restrictions**. Consider the following examples.

EXAMPLE 1. John has at most $15,000 to invest in stocks A and B. It is known that stock A pays 18% dividend while stock B pays only 13%. John's broker advised him not to invest more than $5,000 in stock A since it might not be very secure. John took the advice and also decided that the money invested in stock B should be at least three times that of stock A. How much should John invest in each stock to receive maximum return on dividends?

Solution: If we let x be the amount of money John is to invest in stock A and y be the amount of money to be invested in stock B, then the constraints are:

$$x + y \leq 15,000 \qquad (1)$$
$$x \leq 5,000 \qquad (2)$$
$$x \leq 3y \qquad (3)$$

Also,

$$x \geq 0, \; y \geq 0$$

The dividend from this investment is

$$z = 0.18x + 0.13y \qquad (4)$$

which is to be maximized.

The restrictions given in (1), (2), (3) are called constraints and the linear function (4) is called the objective function. The linear programming problem is to maximize the objective function (4) based on the constraints (1), (2), (3).

EXAMPLE 2. A food company manufactures a new product called DOE which contains two substances I, II. Substance I contains 5 units of nutrient A and 3 units of nutrient B per ounce; Substance II contains 4 units of nutrient A and 5 units of nutrient B per ounce. It is required that the food product DOE must contain at least 70 units of nutrient A and 50 units of nutrient B. The cost for substance I is $1.25 per ounce and that of substance II is $0.75 per ounce. How many ounces of each substance should be used to satisfy the requirements and minimize the cost?

Solution: Again, if we let x be the number of ounces for substance I and y the number of ounces of Substance II used

145

for the product, then the total cost function (objective function) is

$$z = 1.25x + 0.75y$$

which is subject to the constraints

	SUBSTANCE I		SUBSTANCE II		
Nutrient A:	$5x$	$+$	$4y$		≥ 70
Nutrient B:	$3x$	$+$	$5y$		≥ 50

and $x \geq 0$, $y \geq 0$.

The objective is clearly to minimize the cost function

$$z = 1.25x + 0.75y.$$

Thus, a linear programming problem is one to maximize (minimize) a linear function (the objective function)

$$z = a_1 x_1 + a_2 x_2 + \ldots + a_n x_n$$

with x_1, x_2, ..., x_n as variables, subject to the constraints

$$b_{11} x_1 + \ldots + b_{1n} x_n \leq (\geq) \, b_1$$
$$\vdots \qquad\qquad \vdots \qquad\qquad \vdots$$
$$b_{m1} x_1 + \ldots + b_{mn} x_n \leq (\geq) \, b_m$$

It is also assumed that all the variables are non-negative, i.e $x_1 \geq 0$, $x_2 \geq 0$, ..., $x_n \geq 0$.

The set of solutions satisfying the system of linear inequalities is called the set of feasible solutions for the linear programming problem.

In the next two sections, we shall introduce two methods

146

to solve a linear programming problem, first a graphical
(geometrical) method to solve a problem with two variables,
then a simplex method to solve a problem with two or more
variables.

SELF-TEST:

1. A linear programming problem is a problem to maximize or
    minimize a linear function which is called the (a) _____,
    subject to certain restrictions, which are called
    (b) _____. The restrictions are given
    in terms of a set of (c) _____.

    ANS: (a)   objective function
    (b)   constraints
    (c)   inequalities

2. Write the following linear programming problems in terms
    of its objective function and constraints:
    Each type B needs ¼ lb. per flour.
    A baker has at most 50 lbs. of flour to make two
    different kinds of bread. Each bread type A needs
    1/5 lb. of flour. Each bread type B needs
    1/4 lb. of flour. The labor cost for making
    one bread A is 6¢ and that for bread B is 8¢.
    Suppose the baker can make 20¢ profit on each
    bread A and 25¢ on each bread B and he has at
    most $40 to spend, how many loaves of bread of
    each kind should he make in order to maximize
    his profit?

    ANS: Let x and y be the number of bread loaves to
    be made for each kind, the objective function
    is

    $$z = .20x + .25y$$

    subject to the constraints

    $$1/5x + 1/4y \leq 50$$
    $$.06x + .08y \leq 40$$

The problem is to maximize the objective function.

3.2   Graphical Method of Solving Linear Programming Problems

OBJECTIVES:

1.   Introduce the region of feasible solutions of
     the linear programming problem by the graphical
     method.
2.   Introduce the concept that the maximum or
     minimum value of the objective function for
     a linear programming problem with a bounded
     region of feasible solutions always occurs at a
     vertex point (or points).
3.   Demonstrate by examples that a linear pro-
     gramming problem with an unbounded region of
     feasible solutions may or may not have a
     solution. If the solution exists, it must also
     occur at a vertex point.

Consider the following linear programming problem:

EXAMPLE 1.   Maximize the objective function

$$z = 2x + y$$

subject to the linear constraints

$$3x + y \leq 9$$
$$x + y \leq 5$$
$$x \geq 0, \ y \geq 0$$

In Figure 1, the shaded region bounded by the four lines

$$x = 0$$
$$y = 0$$
$$3x + y = 9$$
$$x + y = 5$$

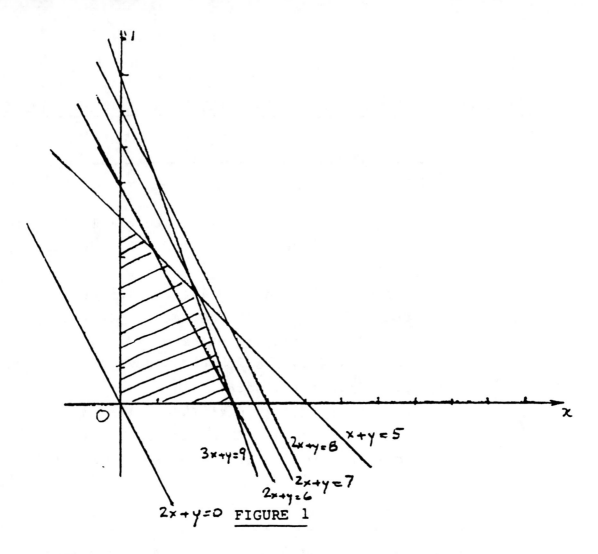

$2x+y=0$ FIGURE 1

clearly represents the solution set of the linear constraints.
That is, any point inside the shaded region or on the boundary
satisfies the four linear inequalities. The region (with
boundary included) is called the region of feasible solutions.
Our aim is to find a point (x,y) in the region of feasible
solutions which maximizes $z = 2x + y$.

Note that for different values of z, the lines representing $z = 2x + y$ are parallel lines since they all have the
same slope -2.   Figure 1 contains the lines $2x + y = 0$,
$2x + y = 6$, $2x + y = 7$, $2x + y = 8$. The line $2x + y = 6$
intersects the shaded region but the line $2x + y = 8$ does
not. This indicates that a maximum solution for $z = 2x + y$
must be one between $z = 6$ and $z = 8$. From Figure 1, it is
clear that the line $z = 2x + y$ must not go beyond the point
$(2, 3)$, which is the point of intersection of the lines
$3x + y = 9$ and $x + y = 5$, since it will be outside of the

150

feasible range. But as long as the line does not reach the point (2,3), we can always increase the value of z = 2x + y by moving it closer to the point (2,3). Hence, we see that the maximum value for z = 2x + y is 7 and it occurs at the point (2,3) which is a vertex (corner) point of the region.

The above example indicated a maximum solution existed and was obtained at a vertex point. This is true if the region of feasible solutions is bounded. For different objective functions, maximum values are obtained at different vertex points. Consider the following example:

EXAMPLE 2: The region of feasible solutions is the same as in Example 1, but the objective function is

$$z = y + 5x$$

From Figure 2, we see that the maximum solution occurs at the vertex point (3,0) and the maximum value is 15.

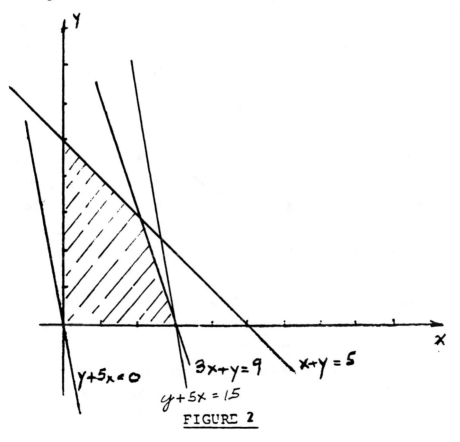

FIGURE 2

151

It is also possible that the maximum value is obtained at more than one vertex point as can be seen from the following example.

EXAMPLE 3. The feasible region is the same as in Example 1 and the objective function is

$$z = 6x + 2y$$

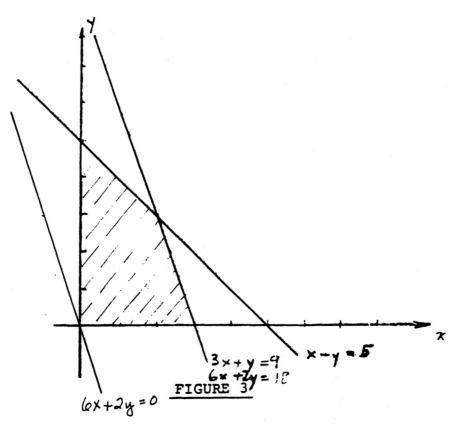

FIGURE 3

From Figure 3 we see that the lines representing the function $z = 6x + 2y$ are **parallel to one of the** boundaries of the region, i.e., $3x + y = 9$. **It is clear that the maximum value for** $z = 3x + y$ **can be obtained at any point on the line segment** between the points **(2,3) and (3,0).**

As indicated before, if the region is bounded, the maximum (or minimum) value of the objective function will occur at one or more vertex points. Figure 4, 5, 6 and 7 demonstrate the occurrence of maximum values of various objective functions $z = f(x,y)$.

152

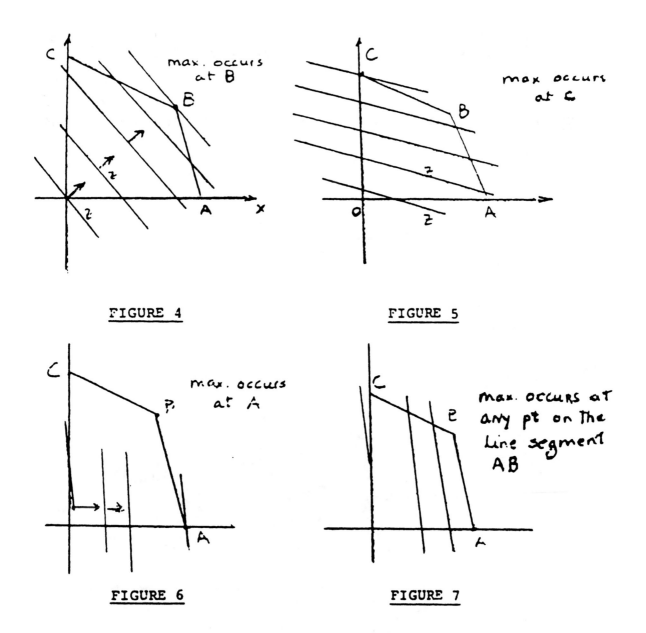

FIGURE 4

FIGURE 5

FIGURE 6

FIGURE 7

If the region is bounded, the maximum (or minimum) value for the objective function occurs at a vertex point. Thus, we can find the maximum (or minimum) value by applying the following simple procedure:

STEP 1. Find all the vertex points of the feasible region (they are the points of intersection of certain pairs of boundary lines of the region).

STEP 2. Evaluate the objective function at each of the vertex points.

The largest value is the maximum and the smallest value is the minimum.

EXAMPLE 4. Find the maximum and minimum values of the objective function

$$z = x + 5y$$

subject to the constraints

$$x \leq 6$$
$$x + y \geq 2$$
$$x + 3y \leq 15$$
$$x \geq 0$$
$$y \geq 0$$

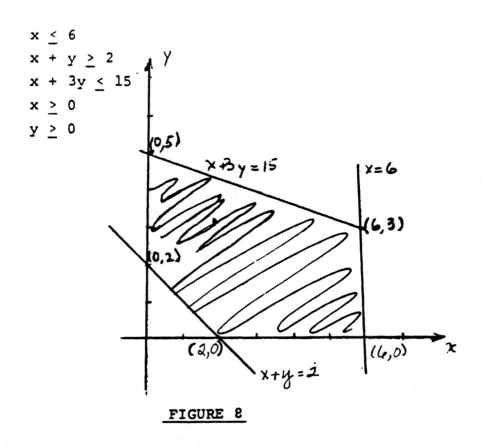

FIGURE 8

Solution. The shaded region in Figure 8 represents the feasible solutions of this problem. Solving each pair of the following equations simultaneously:

$$x = 6$$
$$x + y = 2$$
$$x + 3y = 15$$
$$x = 0$$
$$y = 0$$

154

we obtain all the vertex points of the region as follows:

$(0,2)$, $(2,0)$, $(6,0)$, $(6,3)$, $(0,5)$.

The function value for $z = x + 5y$ at each of the vertex points is given in the following table:

Vertex Point	$z = x+5y$
$(0,2)$	10
$(2,0)$	2
$(6,0)$	6
$(6,3)$	21
$(0,5)$	25

We observe from the table that the maximum value for the objective function is 25 which occurs at the point $(0,5)$ and the minimum value is 2 which occurs at the point $(2,0)$.

SELF-TEST:

1. Maximize the objective function

$$z = 5x + 7y$$

subject to the constraints

$$2x + 3y \leq 6$$
$$x + y \geq 2$$
$$x \geq 0, \ y \geq 0$$

ANS: The maximum value of z is 15 at $x = 3$, $y = 0$.

2. Minimize the objective function

$$z = 2x_1 + 3x_2$$

155

subject to the constraints

$$2x_1 + 3x_2 \leq 12$$
$$3x_1 + x_2 \leq 12$$
$$x_1 + x_2 \geq 2$$
$$x_1 \geq 0, \ x_2 \geq 0$$

ANS: The minimum value of z is 4 obtained at $x_1 = 2$, $x_2 = 0$.

3. A baker has at most 100 pounds of flour to make two different kinds of products A and B. Each of product A needs 0.25 pound of flour and product B requires 0.20 pound of flour. It costs the baker 5¢ to make product A and 12¢ to make product B and he can spend at most $30. If the profit is 10¢ and 15¢ for product A and B respectively, how many of each product should the baker make in order to maximize his profit?

ANS: Let x be the number of product A to be made and y that of B. Then the objective function (the profit) to be maximized is
$$z = 0.10x + 0.15y$$
subject to the constraints
$$0.25x + 0.20y \leq 100$$
$$0.05x + 0.12y \leq 30$$
$$x \geq 0, \ y \geq 0$$
and the solution is z = 48.75, a maximum value obtained at x = 300 and y = 125.

If the region of feasible solutions is not bounded, then it is possible that a maximum value does not exist. Consider the following example:

EXAMPLE 5. Maximize the objective function

$$z = 2x + y$$

156

subject to the linear constraints

$$3x + y \geq 9$$
$$x + y \geq 5$$
$$x \geq 0$$
$$y \geq 0$$

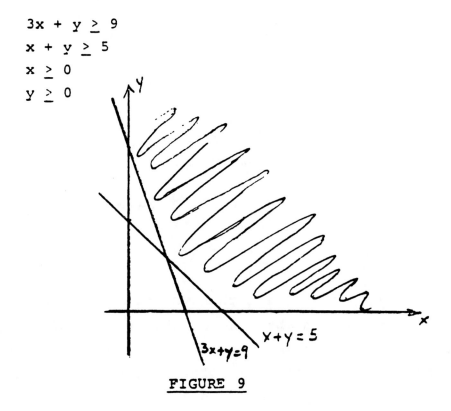

FIGURE 9

From Figure 9, we see that the shaded region is not bounded and the objective function cannot reach a maximum value.

The following is an example of a linear programming problem with unbounded feasible region and a minimum solution.

EXAMPLE 6. Minimize the objective function

$$z = 2x + y$$

subject to the constraints

$$3x + y \geq 9$$
$$x + y \geq 5$$
$$x \geq 0$$
$$y \geq 0$$

157

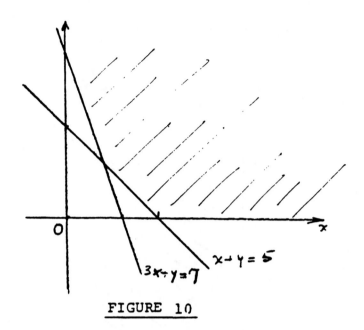

FIGURE 10

From Figure 10, we see that the minimum value for the objective function is obtained at the vertex point (2,3) and the value is 7. Thus we have an example of an unbounded region which has a solution and the solution also occurs at a vertex point. The following theorem is true:

THEOREM 6.1: If an objective function has a solution (minimum or maximum) in the region of feasible solutions, then it must occur at the vertex point.

EXAMPLE 7. Mary needs at least 44 units of vitamin A and 30 units of vitamin B each month. She can obtain the required vitamins from consuming two different foods, I and II. Food I contains 4 and 2 units of vitamin A and B respectively per pound; and food II contains 2 and 3 units of vitamin A and B, per pound respectively. The cost for food I is $1.00 per pound and $0.75 for food II per pound. How many pounds of food I and II should Mary take per month in order to satisfy the vitamin requirements at the lowest cost?

Solution: Assume Mary consumes x pounds of food I and y pounds of food II, then the objective function to be minimized is

158

$$z = 1.0x + 0.75y$$

The constraints are

$$4x + 2y \geq 44$$
$$2x + 3y \geq 30$$
$$x \geq 0, \; y \geq 0$$

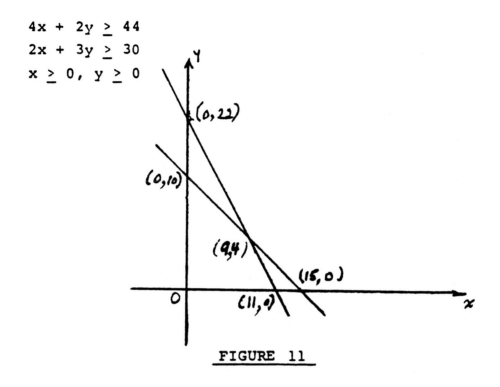

FIGURE 11

Figure 11 indicates the minimum value, obtained at the point (9,4), is 12.

SELF-TEST:

1. **Minimize the objective function**

    $$z = 2x_1 + 3x_2$$

    subject to the constraints

    $$2x_1 + x_2 \geq 10$$
    $$x_1 + 2x_2 \geq 8$$
    $$x_1 \geq 0, \; x_2 \geq 0$$

    ANS: The minimum value of z is 14 obtained at $x_1 = 4$, $x_2 = 2$.

159

2. Maximize the objective function

$$z = 8x + 7y$$

subject to the constraints

$$4x + 3y \geq 24$$
$$3x + 4y \geq 8$$
$$x \geq 0, \ y \geq 0$$

ANS: The region is unbounded; hence, z has no maximum.

EXERCISE 3.2

1. Maximize the objective function

$$z = 4x + 3y$$

subject to

$$x + y \leq 2, \ x \geq 0, \ y \geq 0$$

2. Maximize the objective function

$$z = 5x + 2y$$

subject to

$$2x + 3y \leq 6$$
$$4x + Y \leq 6$$
$$x \geq 0, \ y \geq 0$$

3. Maximize the objective function

$$z = x + 3y$$

subject to

$$2x + 3y \leq 10$$
$$5x + 4y \leq 20$$
$$x \geq 0, \ y \geq 0$$

4. Maximize the objective function

$$z = 15x + 25y$$

subject to

$$3x + 4y \leq 15, \ x \geq 0, \ y \geq 0$$

161

5. Maximize the objective function

$$z = 5x + 10y$$

subject to

$$5x + 5y \leq 30$$
$$2x + 10y \leq 20$$
$$x \geq 0, \ y \geq 0$$

6. Minimize the objective function

$$z = 2x + 3y$$

subject to

$$x + y \geq 2, \ x \geq 0, \ y \geq 0$$

7. Minimize the objective function

$$z = 8x + 7y$$

subject to

$$4x + 3y \geq 16$$
$$3x + 4y \geq 6$$
$$x \geq 0, \ y \geq 0$$

8. Minimize the objective function

$$z = 4x + 5y$$

subject to

$$2x - y \leq 50$$
$$x + y \geq 15$$
$$x \geq 0, \ y \geq 0$$

9. Minimize the objective function

$$z = 2x + 4y$$

subject to

$$x + 2y \geq 10$$
$$3x + y \geq 10$$
$$x \geq 0, \; y \geq 0$$

10. Minimize the objective function

$$z = x + 2y$$

subject to

$$x + y \geq 1$$
$$2x + 4y \geq 3$$
$$x \geq 0, \; y \geq 0$$

11. Maximize and minimize the objective function

$$z = x + 5y$$

subject to

$$x + 4y \leq 12$$
$$x \leq 8$$
$$x + y \geq 2$$
$$x \geq 0, \; y \geq 0$$

12. Maximize and minimize the objective function

$$z = 5x + 2y$$

subject to

163

$$x + y \leq 10$$
$$2x + y \geq 10$$
$$x + 2y \geq 10$$
$$x \geq 0, \ y \geq 0$$

13. Maximize and minimize the objective function

$$z = 15x + 15y$$

subject to

$$x + y \leq 25$$
$$2x - y \leq 20$$
$$-3x + y \leq 5$$
$$x \geq 0, \ y \geq 0$$

14. Maximize and minimize the objective function

$$z = 3x + 4y$$

subject to

$$x + y \leq 5$$
$$2x + y \geq 5$$
$$x + 2y \geq 5$$
$$x \geq 0, \ y \geq 0$$

15. John has $12,000 to invest in stocks A and B. Stock A yields 8% and B yields 10%. It is required that John must invest at least three times as much in A as in B. How much should be invested on each stock in order to receive maximum return?

16. A farmer raises chickens and turkeys. He decides to raise no more than 100 of them altogether and no more than 40 turkeys. He has $120 to spend. It costs $1.50 to raise a chicken and $2.00 for a turkey. The profit from

selling a chicken is $2.00 and that of turkey is $3.00. How many of each should be raised to achieve maximum profit?

17. A baker has at most 200 pounds of flour to make two different kinds of product A and B. Each of product A requires 0.5 pound of flour, and each of product B requires 0.4 pound of flour. The cost to make one of product A is 10¢ and that of B is 24¢. The baker has at most $60 to spend. If the profit is 20¢ and 30¢ for product A and B respectively, how many of each product should the baker make to maximize the profit?

18. A diet contains at least 300 units of vitamins, 1000 units of calories and 400 of minerals. It is known that each of food I contains 2 units of vitamins, 3 units of calories and 1 unit of minerals; and food II contains 1 unit of vitamins, 3 units of calories and 2 units of minerals. If food I costs $0.10 per unit and food II costs $0.05 per unit, how many units of each food should be included so that the minimum nutrition requirements is met and the cost is a minimum?

19. A maufacturing company produces a product A to be shipped to its two warehouses. Warehouse I holds a maximum 100 units of product A and has already 30 in stock. Warehouse II holds a maximum of 80 units and has already 25 on hand. The Company must ship at least 100 units to the two warehouses. The cost to ship a unit to warehouse I is $15 and that to warehouse II is $13. How many units should be shipped to each warehouse in order to maintain minimum shipping costs? What is the minimum cost?

3.3 The Simplex Method

OBJECTIVES:

1. *Discuss the simplex method to solve the maximization problem.*
2. *Consider the minimization problem as the dual problem of maximization.*

The graphical method can be used to solve a linear programming problem with two variables and a small number of constraints. When the number of variables is three or more or the number of constraints is large, the graphical method can no longer be used to solve the problem. We introduce here a matrix method known as the simplex method which was developed by George Dantzig. The basic idea is to change each of the given constraints into an equation and apply the row reduction method to systematically improve the objective function until a solution (minimum or maximum) is found or it can be decided that no solution can be found.

We shall first restrict ourself to the problem of maximization then consider the minimization problem as its dual problem. We now discuss the mthod by considering the example:

<u>EXAMPLE 8.</u> Maximize the objective function

$$z = 2x_1 + x_2$$

subject to contraints

$$3x_1 + x_2 \leq 9$$
$$x_1 + x_2 \leq 5$$

<u>Solution:</u> Each of the constraints is given as an inequality

to change it into an equation, we add a variable which is called a slack variable. Recall that the inequality

$$a \leq c$$

implies

$$c - a \geq 0$$

and the equality

$$a + (c-a) = c$$

can be written as

$$a + b = c \quad \text{where } b \geq 0.$$

Therefore, from the inequality

$$3x_1 + x_2 \leq 9$$

we know there exists a real number $x_3 \geq 0$ such that

$$3x_1 + x_2 + x_3 = 9$$

Also from $x_1 + x_2 \leq 5$ there is a real number $x_4 \geq 0$ such that

$$x_1 + x_2 + x_4 = 5$$

We thus derive two equations based on the two given constraints. In the process of conversion we have introduced two new variables x_3, x_4 and they are called the slack variables.

The problem is to find a solution to the system

$$3x_1 + x_2 + x_3 = 9$$
$$x_1 + x_2 + x_4 = 5$$
$$-2x_1 - x_2 + z = 0$$

167

such that $x_1 \geq 0$, $x_2 \geq 0$, $x_3 \geq 0$, $x_4 \geq 0$ and z is maximum. Note that the last equation is obtained from the objective function. The next step is to write down the augmented matrix (which is called the simplex tableau) of the system which is

$$T = \begin{bmatrix} 3 & 1 & 1 & 0 & 0 & \vdots & 9 \\ 1 & 1 & 0 & 1 & 0 & \vdots & 5 \\ \hline -2 & -1 & 0 & 0 & 1 & \vdots & 0 \end{bmatrix}$$

Notice we also put a dotted line on top of the objective equation. We now look at the last row of the matrix and locate the most negative entry (i.e., the negative entry with largest absolute value in the last row). In our case, -2 is the element to be located since -2, -1 are the only negative entries in the last row, and -2 has larger absolute value 2 than that of -1 which has absolute value 1. The column which has the entry -2 is called the pivot column. In this case, column 1 contains the entry -2, hence, is the pivot column. We now divide each element of the last column by the corresponding element in the pivot column and obtain the following positive numbers:

$$9/3 = 3, \quad 5/1 = 5$$

The entry which has the smallest ratio is called the pivot element. In our example, 3 is less than 5, thus the number 3 which is the (1,1)-entry of the matrix is the pivot element. We then apply the elementary row operations introduced in Chapter 5 to change the pivot element to 1 and all other elements in the pivot column to 0. The resulting matrix is the following:

$$\begin{bmatrix} 1 & 1/3 & 1/3 & 0 & 0 & \vdots & 3 \\ 0 & 2/3 & -1/3 & 1 & 0 & \vdots & 2 \\ \hline 0 & -1/3 & 2/3 & 0 & 1 & \vdots & 6 \end{bmatrix}$$

Notice that the last row is equivalent to the objective function

$$z = 6 + 0 \cdot x_1 + 1/3x_2 - 2/3x_3 + 0 \cdot x_4 - x_5$$

The solution of the above objective function can be improved by increasing the value of x_2. For this purpose, we repeat the same process again by finding the most negative value from the last row and it is -1/3. In fact, it is the only negative entry left in the last row. Column 2 is the pivot column. To select the pivot element from the pivot column we divide each element of the last column by the corresponding element in the pivot column. The ratios are

$$2/\tfrac{2}{3} = 3 \quad \text{and} \quad 3/\tfrac{1}{3} = 9$$

Since 3 < 9, the pivot element is 2/3, the (2,2)-entry of the matrix. Apply row reduction process we have the matrix

$$\begin{bmatrix} 1 & 0 & 1/2 & -1/2 & 0 & \vdots & 2 \\ 0 & 1 & -1/2 & 3/2 & 0 & \vdots & 3 \\ \hdashline 0 & 0 & 1/2 & 1/2 & 1 & \vdots & 7 \end{bmatrix}$$

The last row is equivalent to the objective function

$$z = 7 - 1/2x_3 - 1/2x_4 - x_5$$

Since the coefficients of the variables appeared in the last expression are all negative, it is not possible to improve the value for z, thus the maximum value for z is 7 when $x_3 = x_4 = x_5 = 0$ and the corresponding values for x_1 and x_2 from row 1 and row 2 of the last matrix is $x_1 = 2$, $x_2 = 3$. And this is the solution for this example.

The above example indicated that the procedure of the simplex method used to solve a linear programming problem is as follows:

STEP 1. Change each of the inequalities (constraints) into an equation by introducing a slack variable whose value is non-negative and obtain a system of linear equations with the objective equation included.

STEP 2. Express the system in its augmented matrix form with the objective equation as the last row of the matrix and put dotted lines on top of the last row.

STEP 3. Search for the element in the last row which has the most negative value. Identify the pivot column as the one containing the element a.

STEP 4. Form the ratios which are obtained from dividing each element of the last column by the corresponding element above the dotted line of the pivot column. The element on the pivot column giving the smallest ratio is the pivot element.

STEP 5. Apply the row reduction process (elementary row operations given in Chapter 5) to change the pivot element into 1 and all other elements in the pivot column into zeros.

STEP 6. Repeat STEP 3 to STEP 5 until all the elements of the last row are non-negative. We then obtain the solution by setting all the variables with non-zero coefficients in the last row zero. The value for the other variables can be found from the other rows of the matrix. The solution is the number that appears in the last row and last column.

We now apply the simplex method to solve the following word problem:

EXAMPLE 9. A manufacturing company produces two products P_1 and P_2. Both products require the operation of two machines M_1 and M_1. It is known that P_1 requires 3 minutes of machine time of M_1 and 4 minutes of that of M_2 per unit production. Product P_2 requires 5 minutes of M_1 and 3 minutes of M_2. The total available machine time for M_1 and M_2 is 600 minutes and 400 minutes respectively. If the profit generated from P_1 is \$7 per unit and for P_2 \$6 per unit, how many units should the company produce for each product to obtain maximum profit?

Solution: Let x_1 be the number of units to be produced for P_1 and x_2 that for P_2. The objective function is the profit

$$z = 7x_1 + 6x_2$$

The constraints are those conditions given in terms of the available machine time:

> P_1 requires 3 minutes of M_1 per unit.
> P_2 requires 5 minutes of M_2 per unit.
> Thus, the combined time required for M_1 and M_2 is:

$$3x_1 + 5x_2$$

Since the total available machine time for M_1 is 600, we obtain the first inequality (constaint)

$$3x_1 + 5x_2 \leq 600.$$

Similarly, we can obtain the second constraint based on the time required and available for Machine M_2 which is

$$4x_1 + 3x_2 \leq 400$$

Thus, the problem is to maximize

$$z = 7x_1 + 6x_2$$

subject to

$$3x_1 + 5x_2 \leq 600 \qquad\qquad (1)$$
$$4x_1 + 3x_2 \leq 400 \qquad\qquad (2)$$

We now follow the procedure to solve this problem.

STEP 1. Change inequality (1) into an equation by introducing the slack variable x , thus

$$3x_1 + 5x_2 + x_3 = 600, \; x_3 \geq 0$$

Also, inequality (2) is changed to

$$4x_1 + 3x_2 + x_4 = 400, \text{ with } x_4 \geq 0$$

The system of linear equations is

$$3x_1 + 5x_2 + x_3 = 600$$
$$4x_1 + 3x_2 + x_4 = 400$$
$$-7x_1 - 6x_2 + z = 0$$

STEP 2. The augmented matrix of the equation is

$$\begin{bmatrix} 3 & 5 & 1 & 0 & 0 & | & 600 \\ 4 & 3 & 0 & 1 & 0 & | & 400 \\ \hline -7 & -6 & 0 & 0 & 1 & | & 0 \end{bmatrix}$$

STEP 3. The most negative element in the last row is -7, hence column 1 is the pivot column.

STEP 4. Form the ratio

172

$$600/3 = 200, \quad 400/4 = 100$$

Since 100 is smaller, the pivot element is 4.

STEP 5.   Apply row reduction we have

$$\begin{bmatrix} 0 & 11/4 & 1 & -3/4 & 0 & \vline & 300 \\ 1 & 3/4 & 0 & 1/4 & 0 & \vline & 100 \\ \hline 0 & -3/4 & 0 & 7/4 & 1 & \vline & 700 \end{bmatrix}$$

We repeat STEP 3 - STEP 5.

STEP 3.   The most negative element of the last row is
-3/4 thus the pivot column is column 2.

STEP 4.   The ratios are

$$100/{}_{3/4} = 400/3 \quad \text{and} \quad 300/{}_{11/4} = 1200/11$$

since $1200/11 < 400/3$ the pivot element is 11/4.

STEP 5.   Apply row reduction we have

$$\begin{bmatrix} 0 & 1 & 4/11 & -3/11 & 0 & \vline & 1200/11 \\ 1 & 0 & -3/11 & 5/11 & 0 & \vline & 200/11 \\ \hline 0 & 0 & 3/11 & 17/11 & 1 & \vline & 8600/11 \end{bmatrix}$$

All elements of the last row of above matrix are
non-negative. The procedure should be stopped
and the maximum solution is

$$z = 8600/11$$

which is obtained from $x_2 = 1200/11$, $x_1 = 200/11$.

Hence, the company should produce 109
units of P_2 and 18 units of P_1 to
reach the maximum profit of \$871.

SELF-TEST:

1. Apply the simplex method to maximize the objective function

$$z = 15x_1 + 10x$$

subject to

$2x_1 + x_2 \leq 10$		(a)
$x_1 + 2x_2 \leq 8$		(b)
$x_1 \geq 0, \ x_2 \geq 0$		

Introduce a slack variable x_3 to inequality (a) and obtain
the equation _____.

ANS: $2x_1 + x_2 + x_3 = 10$

Introduce a slack variable x_4 to inequality (b) and obtain
the equation _____.

ANS: $x_1 + 2x_2 + x_4 = 8$

Combine the above equations and the objective function
we have a system of linear equation as follows:

ANS: $2x_1 + x_2 + x_3 = 10$
$x_1 + 2x_2 + x_4 = 8$
$-15x_1 - 10x_2 + z = 0$

The augmented matrix (simplex tableau) of the system is:

ANS: $\begin{bmatrix} 2 & 1 & 1 & 0 & 0 & | & 10 \\ 1 & 2 & 0 & 4 & 0 & | & 8 \\ \hline -15 & -10 & 0 & 0 & 1 & | & 0 \end{bmatrix}$

The most negative element of the last row is _____ .

ANS: -15

Thus the pivot column is column _____ .

ANS: 1

Divide each element above the dotted line by the corresponding element of the pivot column and obtain the two ratios _____ and _____ .

ANS: 8/1 = 8, 10/2 = 5

The smallest ratio is _____ .

ANS: 5

Thus the pivot element is _____ and is the (_,_)-entry.

ANS: 2, (1,1)-entry

Apply the row reduction process. First change the pivot element to 1 and the matrix is changed to:

ANS: $\begin{bmatrix} 1 & 1/2 & 1/2 & 0 & 0 & \vdots & 5 \\ 1 & 2 & 0 & 4 & 0 & \vdots & 8 \\ \hline -15 & -10 & 0 & 0 & 1 & \vdots & 0 \end{bmatrix}$

Change all other elements other than the pivot element on the pivot column and obtain the matrix:

ANS: $\begin{bmatrix} 1 & 1/2 & 1/2 & 0 & 0 & \vdots & 5 \\ 0 & 3/2 & -1/2 & 4 & 0 & \vdots & 3 \\ \hline 0 & -5/2 & 15/2 & 0 & 1 & \vdots & 75 \end{bmatrix}$

Find the most negative element of the last row. It
is _____.

    ANS: -5/2

The pivot column is column _____.

    ANS: 2

Divide each element above the dotted line of the last
column by the corresponding element of the pivot column,
we obtain the ratios _____ and _____.

    ANS: $3/\frac{3}{2} = 2$ and $5/\frac{1}{2} = 10$

The pivot element is _____ and it is the (__,__)
entry.

    ANS: 3/2, (2,2)-entry

Change the pivot element to 1 and obtain the matrix:

$$\text{ANS:} \quad \left[\begin{array}{ccccc|c} 1 & 1/2 & 1/2 & 0 & 0 & 5 \\ 0 & 1 & -1/3 & 8/3 & 0 & 2 \\ \hline 0 & -5/2 & 15/2 & 0 & 1 & 75 \end{array}\right].$$

Change the elements of the pivot **column above and below** the
pivot element and obtain **the matrix:**

$$\text{ANS:} \quad \left[\begin{array}{ccccc|c} 1 & 0 & 2/3 & -4/3 & 0 & 4 \\ 0 & 1 & -1/3 & 8/3 & 0 & 2 \\ \hline 0 & 0 & 20/3 & 20/3 & 1 & 80 \end{array}\right].$$

The solution of this problem is that the maximum value of
z is _____ and obtained at $x_1 = $ _____, $x_2 = $ _____.

max is 80
$$x_1 = 4, \quad x_2 = 2$$

We should point out here that it is possible that the maximization problem may not have a solution. This happens when the elements of the pivot column are either zero or negative.

Consider the example:

EXAMPLE 10. Maximize the objective function

$$z = 2x_1 + 2x_2$$

subject to

$$-2x_1 + 2x_2 \leq 4$$
$$3x_1 - 3x_2 \leq 6$$

Adding the slack variables x_3 and x_4 we have

$$-2x_1 + 2x_2 + x_3 = 4$$
$$3x_1 - 3x_2 + x_4 = 6$$

The augmented matrix is

$$\left[\begin{array}{ccccc|c} -2 & 2 & 1 & 0 & 0 & 4 \\ 3 & -3 & 0 & 1 & 0 & 6 \\ \hline -2 & -2 & 0 & 0 & 1 & 0 \end{array}\right]$$

We choose the first column as the pivot column and 3 is the pivot element. After applying the row reduction process, we obtain the matrix

177

$$\begin{bmatrix} 0 & 0 & 1 & 2/3 & 0 & \vdots & 8 \\ 1 & -1 & 0 & 1/3 & 0 & \vdots & 2 \\ \hline 0 & -4 & 0 & 2/3 & 1 & \vdots & 4 \end{bmatrix}$$

The only non-negative element of the last row is -4, thus the pivot column should be column 2. However, all the elements in the second column are zero or negative. We conclude that the problem has no solution or sometimes we say it has unbounded solution.

SELF-TEST:

If the pivot column of the matrix contains elements which are
_____ or _____ then the maximization problem has
no solution.

ANS: zero or negative

All the examples given so far on maximization problems have constraints expressed in terms of the inequality less than or equal to ($\leq$) which allows us to add a slack non-negative variable to change an inequality into an equation form. In case some of the inequalities go to the other direction, i.e., greater than or equal to ($\geq$) we should subtract a slack variable instead of adding. For example, the inequality

$$2x_1 + 3x_2 + x_3 \geq 5$$

can be changed to

$$2x_1 + 3x_2 + x_3 - x_4 = 5$$

for some non-negative variable x_4.

EXAMPLE 11. Maximize the objective function

$$z = 3x_1 + 2x_2$$

subject to the constraints

$$2x_1 + x_2 \leq 8 \qquad\qquad (3)$$
$$x_1 + 2x_2 \geq 8 \qquad\qquad (4)$$
$$x_1 \geq 0, \ x_2 \geq 0$$

Add to the left side of inequality (3) the slack variable x_3 and we have

$$2x_1 + x_2 + x_3 = 8$$

Subtract from the left side of inequality (4) the slack variable x_4 and we have

$$x_1 + 2x_2 - x_4 = 8$$

Combine the above two equations and the objective equation and we have a system of linear equations with the following augmented matrix:

$$\begin{bmatrix} 2 & 1 & 1 & 0 & 0 & | & 8 \\ 1 & 2 & 0 & -1 & 0 & | & 8 \\ \hline -3 & -2 & 0 & 0 & 1 & | & 0 \end{bmatrix}$$

Column 1 should be the pivot column and the element 2 is the pivot element. We obtain the following matrix after row reduction.

$$\begin{bmatrix} 1 & 1/2 & 1/2 & 0 & 0 & | & 4 \\ 0 & 3/2 & -1/2 & -1 & 0 & | & 4 \\ \hline 0 & -1/2 & 3/2 & 0 & 1 & | & 12 \end{bmatrix}$$

Repeat the process and we have the matrix:

$$\begin{bmatrix} 1 & 0 & 2/3 & 1/3 & 0 & \vdots & 8/3 \\ 0 & 1 & -1/3 & -2/3 & 0 & \vdots & 8/3 \\ \hline 0 & 0 & 4/3 & -1/3 & 1 & \vdots & 40/3 \end{bmatrix}$$

$$\begin{bmatrix} 3 & 0 & 2 & 1 & 0 & \vdots & 8 \\ 2 & 1 & 1 & 0 & 0 & \vdots & 8 \\ \hline 1 & 0 & 2 & 0 & 1 & \vdots & 16 \end{bmatrix}$$

The the maximum value for z is 16 obtained at $x_1 = 0$, $x_2 = 8$.

SELF-TEST:

1. Add a non-negative slack variable x_3 to change the following inequality into an equation form:

$$6x_1 + 7x_2 \geq 25$$

ANS: $6x_1 + 7x_2 - x_3 = 25$
$x_3 \geq 0$

2. Maximize the objective function

$$z = 15x_1 - 10x_2 + 5x_3$$

subject to

$$x_1 + 2x_2 + 4x_3 \leq 40 \qquad \text{(a)}$$
$$x_1 + 6x_2 + x_3 \geq 60 \qquad \text{(b)}$$
$$x_1 \geq 0, \ x_2 \geq 0, \ x_3 \geq 0$$

add a slack variable x_4 and change inequality (a) into an equation.

ANS: $x_1 + 2x_2 + 4x_3 + x_4 = 40$, $x_4 \geq 0$

add a slack variable x_5 and change inequality (b) into

180

an equation.

ANS: $x_1 + 6x_2 + 3x_3 - x_5 = 60$, $x_5 \geq 0$

The augmented matrix of the system is:

ANS:
$$\left[\begin{array}{cccccc|c} 1 & 2 & 4 & 1 & 0 & 0 & 40 \\ 1 & 6 & 3 & 0 & -1 & 0 & 60 \\ \hline -15 & 10 & -5 & 0 & 0 & 1 & 0 \end{array}\right]$$

The solution is:

the maximum value for z is _____ at x_1 = _____, x_2 = _____, x_3 = _____

ANS: max z = 600, x_1 = 40, x_2 = 0, x_3 = 0

Minimization Problem:

So far we have only discussed the simplex method to solve the maximization problem. As we shall see the method can also be used to solve the minimization problem. In fact, a minimization problem can be changed to a maximum problem. The solution for the given problem, if it exists, is the same as the corresponding maximum problem. This is the reason why we call each problem the dual problem of each other. We demonstrate the procedure by considering the following example:

EXAMPLE 12. Minimize the objective function

$$z = 6x_1 + 3x_2$$

subject to the constraints

$$x_1 + x_2 \geq 2 \tag{5}$$
$$2x_1 + 6x_2 \geq 6 \tag{6}$$

181

Write them first in the following matrix form:

$$
\begin{bmatrix}
1 & 1 & \vdots & 2 \\
2 & 6 & \vdots & 6 \\
\hline
6 & 3 & \vdots & 0
\end{bmatrix}
\begin{array}{l}
\leftarrow \text{inequality (5)} \\
\leftarrow \text{inequality (6)} \\
\\
\leftarrow \text{objective function}
\end{array}
$$

The transpose of the matrix is

$$
\begin{bmatrix}
1 & 2 & \vdots & 6 \\
1 & 6 & \vdots & 3 \\
\hline
2 & 6 & \vdots & 0
\end{bmatrix}
$$

The corresponding dual problem can be stated as:

maximize the objective function

$$w = 2y_1 + 6y_2$$

subject to the constraints

$$y_1 + 2y_2 \leq 6$$
$$y_1 + 6y_2 \leq 3$$

The maximum solution for w is the same as the minimum solution for z. Carry out the simplex method to solve the maximization problem as follows:

$$
\begin{bmatrix}
1 & 2 & 1 & 0 & 0 & \vdots & 6 \\
1 & 6 & 0 & 1 & 0 & \vdots & 3 \\
\hline
-2 & -6 & 0 & 0 & 1 & \vdots & 0
\end{bmatrix}
$$

$$\begin{bmatrix} 2/3 & 0 & 1 & -1/3 & 0 & \vdots & 5 \\ 1/6 & 1 & 0 & 1/6 & 0 & \vdots & 1/2 \\ \hline -1 & 0 & 0 & 1 & 1 & \vdots & 3 \end{bmatrix}$$

$$\begin{bmatrix} 0 & -4 & 1 & -1 & 0 & \vdots & 3 \\ 1 & 6 & 0 & 1 & 0 & \vdots & 3 \\ \hline 0 & 6 & 0 & 2 & 1 & \vdots & 6 \end{bmatrix}$$

The maximum value for w is 6 which is the minimum value for z. Since we were working on the transpose of the original matrix, the third column corresponds to x_1 and the 4th corresponds to x_2. Therefore, the minimum value for z obtained at $x_1 = 0$, $x_2 = 2$ is 6.

SELF-TEST:

Minimize the objective function

$$z = 5x_1 + 2x_2$$

subject to

$2x_1 + 3x_2 \geq 6$ (a)

$2x_1 + x_2 \geq 7$ (b)

$x_1 \geq 0$, $x_2 \geq 0$

Write the inequalties (a), (b) and the objective function in matrix form:

ANS:
$$\begin{bmatrix} 2 & 3 & \vdots & 6 \\ 2 & 1 & \vdots & 7 \\ \hline 5 & 2 & \vdots & 0 \end{bmatrix}$$

The transpose of the above matrix is:

ANS:
$$\left[\begin{array}{cc|c} 2 & 2 & 5 \\ 3 & 1 & 2 \\ \hline 6 & 7 & 0 \end{array}\right]$$

The corresponding maximization problem is to maximize the objective function.

ANS: $w = 6y_1 + 7y_2$

subject to constraints

(a) _____

(b) _____

ANS: (a) $2y_1 + 2y_2 \leq 5$

(b) $3y_1 + y_2 \leq 2$

Solve the maximization problem and the final matrix is of the form:

ANS:
$$\left[\begin{array}{ccccc|c} -4 & 0 & 1 & -2 & 0 & 1 \\ 3 & 1 & 0 & 1 & 0 & 2 \\ \hline 15 & 0 & 0 & 7 & 1 & 14 \end{array}\right]$$

The maximum value for w which is the same as the minimum value for z is _____.

ANS: 14

z reaches its minimum value at $x_1 =$ _____, $x_2 =$ _____ which corresponds to the _____ and _____ column of the last matrix respectively.

ANS: $x_1 = 0$, $x_2 = 7$, 3rd column, 4th column

EXERCISES 3.3

Apply the simplex method to solve each of the following problems:

1. Maximize the objective function

$$z = x + 2y$$

subject to

$$x + y \leq 4$$
$$x + 4y \leq 7$$
$$x \geq 0, \ y \geq 0$$

2. Maximize the objective function

$$z = 5x + 2y$$

subject to

$$2x + 3y \leq 6$$
$$4x + y \leq 6$$
$$x \geq 0, \ y \geq 0$$

3. Maximize the objective function

$$z = 3x - 2y + z$$

subject to

$$x + y + 2z \leq 20$$
$$x + 2y + z \geq 30$$
$$x \geq 0, \ y \geq 0, \ z \geq 0$$

4. Maximize the objective function

$$z = 4x + 6y + 5z$$

subject to

$$2x + 2y + z \leq 4$$
$$x - 4y + 3z \leq 6$$
$$x \geq 0, \ y \geq 0, \ z \geq 0$$

5. Maximize the objective function

$$z = x + 2y + z + 5w$$

subject to

$$x + 2y + z + w \leq 25$$
$$3x + y + 2z + w \leq 50$$
$$x \geq 0, \ y \geq 0, \ z \geq 0, \ w \geq 0$$

6. Minimize the objective function

$$z = 2x + 3y$$

subject to

$$2x + y \geq 2$$
$$x \geq 0, \ y \geq 0$$

7. Minimize the objective function

$$z = 2x + y$$

subject to

$$x + y \leq 8$$
$$x + 2y \geq 5$$
$$x \geq 0, \ y \geq 0$$

8. Minimize the objective function

$$z = 3x + y + z$$

subject to

$$x + y + z \geq 5$$
$$2x + y \geq 4$$
$$x \geq 0, \; y \geq 0, \; z \geq 0$$

9. Minimize the objective function

$$z = x + 2y + z$$

subject to

$$x - 3y + 4z \geq 12$$
$$3x + y + 2z \geq 10$$
$$x - y - z \geq -8$$
$$x \geq 0, \; y \geq 0, \; z \geq 0$$

10. Minimize the objective function

$$z = 5x_1 + 4x_2 + 5x_3 + 6x_4$$

subject to

$$x_1 + x_2 \leq 12$$
$$x_3 + x_4 \leq 25$$
$$x_1 + x_3 \geq 20$$
$$x_2 + x_4 \geq 15$$
$$x_1 \geq 0, \; x_2 \geq 0, \; x_3 \geq 0, \; x_4 \geq 0$$

3.4 SOME APPLICATIONS OF LINEAR PROGRAMMING

OBJECTIVES:

To introduce some applications of linear programming such as the transportation problem, the assignment problem and the traveling salesperson problem.

Linear programming has a wide range of applications in business and economics as well as in many other fields. We shall mention three such applications in this section:

1. **The Transportation Problem.** The original simplex method was used to solve this kind of problem. It has something to do with minimizing the shipping cost. We will demonstrate the problem by considering the following example:

A manufacturing company produces a product at two factories and has three distributing warehouses. The minimum production amount of each factory is given in the following table:

FACTORY I	800 units
FACTORY II	600 units

The minimum demand from each warehouse is given as:

WAREHOUSE I	250 units
WAREHOUSE II	150 units
WAREHOUSE III	400 units

The shipping cost per unit from each of the two factories to each of the three warehouses is as follows:

	WAREHOUSE I	WAREHOUSE II	WAREHOUSE III
FACTORY I	2	3	5
FACTORY II	4	2	3

The problem is to determine the total number of units to be shipped from each factory to each warehouse so that the demands of the warehouses are met and that the total shipping cost is minimum.

We shall let the total number of units shipped from Factory i to Warehouse j be x_{ij}, thus x_{11} is the number of units shipped from Factory I to Warehouse I, x_{12} is that from Factory I to Warehouse II, etc. Based on the table of shipping cost per unit, we obtain the total shipping cost as follows:

$$y = 2x_{11} + 3x_{12} + 5x_{13} + 4x_{21} + 2x_{22} + 3x_{23}.$$

and this is our objective function. We are to minimize the above objective function based on the following constraints:

$$x_{11} + x_{12} + x_{13} \geq 800 \quad \text{(minimum production of Factory I)}$$
$$x_{21} + x_{22} + x_{23} \geq 600 \quad \text{(minimum production of Factory II)}$$
$$x_{11} + x_{21} \geq 250 \quad \text{(minimum demand of Warehouse I)}$$
$$x_{12} + x_{22} \geq 150 \quad \text{(minimum demand of Warehouse II)}$$
$$x_{13} + x_{23} \geq 400 \quad \text{(minimum demand of Warehouse III)}$$

also $x_{ij} \geq 0$ for all i, j.

We now apply the simplex method to solve this transportation problem.

The augmented matrix for the minimum problem is:

$$\begin{bmatrix} 1 & 1 & 1 & 0 & 0 & 0 & \vdots & 800 \\ 0 & 0 & 0 & 1 & 1 & 1 & \vdots & 600 \\ 1 & 0 & 0 & 1 & 0 & 0 & \vdots & 250 \\ 0 & 1 & 0 & 0 & 1 & 0 & \vdots & 150 \\ 0 & 0 & 1 & 0 & 0 & 1 & \vdots & 400 \\ \hline 2 & 3 & 5 & 4 & 2 & 3 & \vdots & 0 \end{bmatrix}$$

The transpose of the matrix is

$$\begin{bmatrix} 1 & 0 & 1 & 0 & 0 & \vdots & 2 \\ 1 & 0 & 0 & 1 & 0 & \vdots & 3 \\ 1 & 0 & 0 & 0 & 1 & \vdots & 5 \\ 0 & 1 & 1 & 0 & 0 & \vdots & 4 \\ 0 & 1 & 0 & 1 & 0 & \vdots & 2 \\ 0 & 1 & 0 & 0 & 1 & \vdots & 3 \\ \hline 800 & 600 & 250 & 150 & 400 & \vdots & 0 \end{bmatrix}$$

The corresponding maximum problem is to maximize:

$$z = 800y_1 + 600y_2 + 250y_3 + 150y_4 + 400y_5$$

subject to the constraints

$y_1 + y_3 \le 2$

$y_1 + y_4 \le 3$

$y_1 + y_5 \le 5$

$y_2 + y_3 \le 4$

$y_2 + y_4 \le 2$

$y_2 + y_5 \le 3$

and $y_1 \ge 0, \ y_2 \ge 0, \ y_3 \ge 0, \ y_4 \ge 0, \ y_5 \ge 0$

Carry out the simplex method to solve the maximum problem as follows:

$$
\left[
\begin{array}{cccccccccccc|c}
1 & 0 & 1 & 0 & 0 & 1 & 0 & 0 & 0 & 0 & 0 & 0 & 2 \\
1 & 0 & 0 & 1 & 0 & 0 & 1 & 0 & 0 & 0 & 0 & 0 & 3 \\
1 & 0 & 0 & 0 & 1 & 0 & 0 & 1 & 0 & 0 & 0 & 0 & 5 \\
0 & 1 & 1 & 0 & 0 & 0 & 0 & 0 & 1 & 0 & 0 & 0 & 4 \\
0 & 1 & 0 & 1 & 0 & 0 & 0 & 0 & 0 & 1 & 0 & 0 & 2 \\
0 & 1 & 0 & 0 & 1 & 0 & 0 & 0 & 0 & 0 & 1 & 0 & 3 \\
\hline
-800 & -600 & -250 & -150 & -400 & 0 & 0 & 0 & 0 & 0 & 0 & 1 & 0
\end{array}
\right]
$$

$$
\left[
\begin{array}{cccccccccccc|c}
1 & 0 & 1 & 0 & 0 & 1 & 0 & 0 & 0 & 0 & 0 & 0 & 2 \\
0 & 0 & -1 & 1 & 0 & -1 & 1 & 0 & 0 & 0 & 0 & 0 & 1 \\
0 & 0 & -1 & 0 & 1 & -1 & 0 & 1 & 0 & 0 & 0 & 0 & 3 \\
0 & 1 & 1 & 0 & 0 & 0 & 0 & 0 & 1 & 0 & 0 & 0 & 4 \\
0 & 1 & 0 & 1 & 0 & 0 & 0 & 0 & 0 & 1 & 0 & 0 & 2 \\
0 & 1 & 0 & 0 & 1 & 0 & 0 & 0 & 0 & 0 & 1 & 0 & 3 \\
\hline
0 & -600 & 550 & -150 & -400 & 800 & 0 & 0 & 0 & 0 & 0 & 1 & 1600
\end{array}
\right]
$$

$$
\left[
\begin{array}{cccccccccccc|c}
1 & 0 & 1 & 0 & 0 & 1 & 0 & 0 & 0 & 0 & 0 & 0 & 2 \\
0 & 0 & -1 & 1 & 0 & -1 & 1 & 0 & 0 & 0 & 0 & 0 & 1 \\
0 & 0 & -1 & 0 & 1 & -1 & 0 & 1 & 0 & 0 & 0 & 0 & 3 \\
0 & 0 & 1 & -1 & 0 & 0 & 0 & 0 & 1 & -1 & 0 & 0 & 2 \\
0 & 1 & 0 & 1 & 0 & 0 & 0 & 0 & 0 & 1 & 0 & 0 & 2 \\
0 & 0 & 0 & -1 & 1 & 0 & 0 & 0 & 0 & -1 & 1 & 0 & 1 \\
\hline
0 & 0 & 550 & 450 & -400 & 800 & 0 & 0 & 0 & 600 & 0 & 1 & 2800
\end{array}
\right]
$$

$$
\left[
\begin{array}{cccccccccccc|c}
1 & 0 & 1 & 0 & 0 & 1 & 0 & 0 & 0 & 0 & 0 & 0 & 2 \\
0 & 0 & -1 & 1 & 0 & -1 & 1 & 0 & 0 & 0 & 0 & 0 & 1 \\
0 & 0 & -1 & 1 & 0 & -1 & 0 & 1 & 0 & 1 & -1 & 0 & 2 \\
0 & 0 & 1 & -1 & 0 & 0 & 0 & 0 & 1 & -1 & 0 & 0 & 2 \\
0 & 1 & 0 & 1 & 0 & 0 & 0 & 0 & 0 & 1 & 0 & 0 & 2 \\
0 & 0 & 0 & -1 & 1 & 0 & 0 & 0 & 0 & -1 & 1 & 0 & 1 \\
\hline
0 & 0 & 550 & 50 & 0 & 900 & 0 & 0 & 0 & 200 & 400 & 1 & 3200
\end{array}
\right]
$$

191

Therefore, the minimum cost is 3,200 which is obtained at
$x_{11} = 800$, $x_{12} = 0$, $x_{13} = 0$, $x_{21} = 0$, $x_{22} = 200$, $x_{23} = 400$.

The Assignment Problem: This problem is to assign each of the
given number of people to a job from the same number of jobs
available so that the cost function is minimum. We may state
the problem as follows:

> There are exactly n jobs to be assigned to n people. It
> is assumed that each person is assigned to exactly one
> job and each job is assigned to exactly one person. We
> are given the costs of assigning any person to any job.
> The problem is to make the assignment so that the cost is
> minimum.

We may formulate the problem as below:

> Suppose that the cost for assigning the i-th person
> to the j-th job is denoted as c_{ij} and we introduce
> the variable x_{ij} as that

$$x_{ij} = \begin{cases} 1 \text{ if the } i\text{-th person is assigned to the } j\text{-th} \\ 0 \text{ otherwise} \end{cases}$$

The total cost function is

$$z = c_{11}x_{11} + c_{12}x_{12} + \ldots + c_{nn}x_{nn}$$

which is to be minimized based on the constraints

$$x_{1j} + x_{2j} + \ldots + x_{nj} = 1 \quad \text{for } j = 1, 2, \ldots, n$$
$$\text{and} \quad x_{i1} + x_{i2} + \ldots + x_{in} = 1 \quad \text{for } i = 1, 2, \ldots, n$$

We remark here that it is sufficient to give only the cost matrix

$$C = [c_{ij}]_{n \times n}$$

for an assignment problem.

3. The Traveling Salesperson Problem: This is a well-known problem and its mathematical formulation is very similar to the assignment problem. We state the problem as follows:

A traveling salesperson is to visit each of the n cities once and exactly once. Assume he/she starts from City 1, he/she must return to City 1. The distance between any two cities is given. The problem to choose a route so that the total distance of traveling is minimum.

To formulate the problem as one in terms of linear programming problem, we first introduce the variables x_{ij} as

$$x_{ij} = \begin{cases} 1 & \text{if the person travels from city } i \text{ to city } j \\ 0 & \text{otherwise.} \end{cases}$$

Assume that the distance from city i to city j is d_{ij}, the total distance to travel is:

$$z = d_{11}x_{11} + d_{12}x_{12} + \ldots + d_{nn}x_{nn}$$

which is to be minimized. The constraints are

$$x_{1j} + x_{2j} + \ldots + x_{nj} = 1, \quad j = 1, 2, \ldots, n$$

(This is obtained from the condition that each city must be visited exactly once).

$$x_{i1} + x_{i2} + \ldots + x_{in} = 1, \quad i = 1, 2, \ldots, n$$

(This is obtained from the fact that exactly one city should be reached from a previously visited city.)

EXERCISES 3.4

Apply the simplex method to solve the following transportation problem:

The shipping cost from each factory to each warehouse is given by the following matrix:

	WAREHOUSE I	WAREHOUSE II	WAREHOUSE III
FACTORY I	1	2	4
FACTORY II	2	2	3

The supply from each factory is given by the column matrix

FACTORY I	500
FACTORY II	600

The demand of each warehouse is given as

WAREHOUSE I	150
WAREHOUSE II	100
WAREHOUSE III	200

Determine the number of units to be shipped from each factory to each warehouse so that the demand is met and the shipping cost is minimum.

REVIEW OF CHAPTER THREE

1. A linear programming problem is to maximize or to minimize a linear function subject to certain restrictions. The linear function is called the _____ and the restrictions are called the _____.

2. The set of solutions satisfying the set of linear inequalities of a linear programming problem is called the set of _____ solutions.

3. The graphical method is used to solve a linear programming problem with _____ variables.

4. The maximum or minimum value of the objective function for a linear programming problem with bounded region of feasible solutions always occurs at a_____ point.

5. If the region of feasible solutions is unbounded, then the linear programming problem _____ or _____ have a solution. If it has a solution it must occur at a _____ point.

6. To apply the simplex method to solve a linear programming problem, we introduce new variables to change each inequality into equations. The variables introduced are called the _____.

7. In the simplex method, the matrix obtained from the system of equations is called _____ or _____.

8. What are the pivot column and pivot element of the augmented matrix described in the simplex method.

9. The procedure for the simplex method should be terminated when all the elements of the last row are all _____.

10. The simplex method used to solve a maximization problem might not yield a solution and this happens when the elements of the pivot column are _____.

11. The simplex method can also be used to solve the minimization problem which is called the _____ problem of the maximization problem.

12. To solve a minimization problem by using the simplex method we first form the augmented matrix of the system, then find its _____ which gives us the corresponding _____ problem.

SAMPLE TEST: CHAPTER THREE

1. Apply the graphical method to solve the following linear programming problem:

   (a) Maximize the objective function

   $$z = 2x + 5y$$

   subject to the constraints

   $$3x + 2y \leq 6$$
   $$x + 3y \leq 5$$
   $$x \geq 0, \ y \geq 0$$

   (b) Minimize the objective function

   $$z = 3x + 6y$$

   subject to the constraints

   $$x + 2y \geq 10$$
   $$2x + 3y \geq 15$$
   $$x \geq 0, \ y \geq 0$$

   (c) ABC Toy Company makes two different kinds of toys, model A and model B. Each model A requires 2 hours of machine I and 2 hours of machine II and each of model B requires 4 hours of machine I and 2 hours of machine II. The company has 3 machine I's and 2 machine II's. Each of the machines work 40 hours per week. Suppose each model A makes $4 profit and each model B makes $5. How many of each model should be made in order to obtain maximum profit?

2. Apply the simplex method to solve the following linear

programming problem

(a) Maximize the objective function

$$z = x + 2y + 3z$$

subject to the constraints

$$2x + y + z \le 25$$
$$2x + 3y + 3z \le 30$$
$$x \ge 0, \; y \ge C, \; z \ge 0$$

(b) Minimize the objective function

$$z = x + 6y + 8z$$

subject to the constraints

$$3x + 3y + 5z \ge 20$$
$$2x + y + 3z \ge 9$$
$$5x + 6y + 2z \ge 30$$
$$x \ge 0, \; y \ge 0, \; z \ge 0$$

COMPUTER APPLICATIONS: CHAPTER THREE

1. Write a BASIC program to carry out the graphical solution
    for a linear programming problem with two variables.

2. Write a BASIC program to carry out the simplex method
    for a maximization problem.

CHAPTER FOUR

PROBABILITY

The study of probability began as a response to mathematicians such as Pascal and Fermat examining games of chance. Even today many good examples for the theory of probability can be drawn from experiments such as tossing coins, rolling dice, and dealing card hands. We will consider the theory of finite probability spaces, i.e., experiments in which the possible number of outcomes is finite. But before we begin the study of probability theory it will be necessary to develop skills from combinatorial analysis. These skills will enable us to "count" the finite outcomes that will be associated with a particular experiment.

4.1 FUNDAMENTAL PRINCIPLE OF COUNTING

OBJECTIVES:

1. *To apply the Fundamental Principle of Counting.*
2. *To count using tree graphs.*

In many of the experiments that we will consider it will be necessary to know the number of possible outcomes. We can list all of these outcomes with a tree graph or apply the Fundamental Principle of Counting (FPC) to find this number.

EXAMPLE 1. Suppose a die is rolled and then a coin is tossed. How many possible outcomes could there be to this experiment?

Solution: The answer can be illustrated with a tree graph. When we roll the die there are _six_ possible outcomes {1, 2, 3, 4, 5, 6}. Then given these six, there are two possible outcomes for tossing the coin {Head, Tail}. Thus the tree graph:

Notice that following the various branches of the tree we can find 12 paths from the starting position, i,e., {1H,1T,2H,2T, 3H,3T,4H,4T,5H,5T,6H,6T}. Hence there are 12 possible outcomes.

EXAMPLE 2. Suppose a coin is tossed 3 times. How many outcomes could there be in this experiment?

Solution: On each toss there are two possible outcomes {Head, Tail}. Hence the tree graph below lists these possibilities.

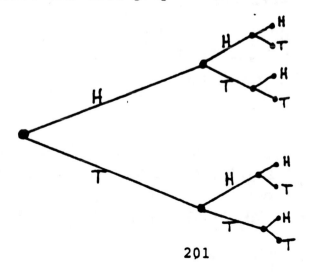

201

Following the paths of the tree we have 8 outcomes {HHH,HHT, HTH,HTT,THH,THT,TTH,TTT}.

The above two examples also could be solved using the Fundamental Principle of Counting.

> FPC: If one experiment has M outcomes and a second experiment has N outcomes, then the two experiments can be performed consectively in MN ways.

Hence in Example 1, we have 6 possible outcomes for the first experiment and 2 possible outcomes for the second, giving a total of (6)(2) = 12 total outcomes.

In Example 2, we have 2 possible outcomes for the first, 2 possible outcomes for the second and 2 possible outcomes for the third. Hence (2)(2)(2) = 8 total possible outcomes.

EXAMPLE 3. How many different numbers can be formed using the digits {1,2,3} if no repetition of digits is permitted?

Solution: Using a tree graph, we have

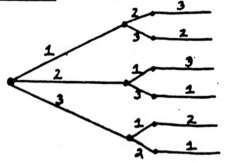

Notice in counting the paths that there are 6 possible outcomes {123,132,213,231,312, 321}.

We could also approach the problem using FPC. There are 3 choices for the 1st digit, 2 choices for the 2nd digit and 1 choice for the 3rd digit. Hence, (3)(2)(1) = 6 possible choices for numbers.

EXAMPLE 4. A license plate consists of two letters followed by 2 digits from {0,1,2,3,4,5,6,7,8,9}. How many possible license plates are there?

Solution: By FPC there are (26)(26)(10)(10) = 67600 plates.

EXERCISES 4.1

1. A license plate consists of 3 letters followed by 3 integers from 0 to 9. How many different license plates are possible?

2. A license plate consists of 2 letters followed by 4 integers from 0 to 9. How many different license plates are possible?

3. If 6 runners compete in a race, in how many ways can the first 3 places be won, if there are no ties?

4. If 7 runners compete in a race, in how many ways can the first 3 places be won, if there are no ties?

5. An experiment consists in drawing from a carton that contains a green, a yellow, and an orange ball; then tossing a coin. Use a tree graph to represent and determine the number of possible outcomes.

6. An experiment consists in rolling a die and then tossing a coin twice. Use a tree graph to represent and determine the number of possible outcomes.

7. Use a tree graph to represent and determine the number of possible two digit numbers which could be formed from the digits 1, 2, 3, 4 if no repetition of digits is allowed.

8. Use a tree graph to represent and determine the number of possible two digit numbers which could be formed from the digits 1, 2, 3, 4 if repetition of digits is allowed.

9. If there are 20 members in a certain club, how many ways can a president, vice-president and treasurer be chosen if no person may serve in more than one office?

10. A restaurant offers a choice of 4 appetizers, 7 main dishes, 8 desserts and 5 beverages. How many complete dinners are available?

11. There are exactly seven roads from Vinton to Needleton and exactly 4 roads from Needleton to Vidor. How many ways may a person make a round trip from Vinton to Vidor if he wishes never to travel on the same road more than once?

12. If a coin is tossed 2 times, how many possible sequences could result?

13. If a coin is tossed 5 times, how many possible sequences could result?

14. If a red die is rolled and then a green die is rolled how many possible outcomes could result?

15. Represent the experiment in Problem 14 with a tree graph.

16. If a card is chosen from an ordinary deck of cards, and then a die is rolled, how many possible outcomes could result?

17. If a card is chosen from an ordinary deck of cards and its suit noted, and then a die is rolled and the parity (odd or even) of the result is noted, how many possible outcomes could there be to this experiment?

18. Represent Problem 17 with a tree graph.

19. A "true or false" quiz consists of 10 questions. How many possible ways might the 10 questions be answered?

20. If a 20 question multiple choice exam has four choices for each question, in how many possible ways could a

student submit a set of answers for this exam?

21. If a signal is to be sent using two flags and each flag may be moved into 6 possible positions, how many signals are possible?

22. In how many ways can 6 books be arranged on a shelf if 2 of them are a set and must be together?

23. In how many ways can ten children line up in single file if two of the children must stand next to each other?

24. How many ways can 7 books be arranged on a shelf if 3 of them are a set and must remain together?

25. How many four-letter arrangements can be made using the letters {a,b,c,d,e} if repetitions are allowed but no letter may follow itself?

4.2 PERMUTATIONS AND COMBINATIONS

OBJECTIVES:

    1. To develop formulae for permutations and combinations.

Let us now consider two special types of problems that arise in counting outcomes of an experiment.

<u>PROBLEM 1.</u> Suppose we have five people and we need to elect a distinct president, vice-president and treasurer. How many possible outcomes could there be?

<u>PROBLEM 2.</u> Suppose we have five people and we wish to choose a committee of three. How many different committees could be formed?

Both of these problems are related and hence we will consider their solutions together. Suppose the five people are represented by {A,B,C,D,E}. Then by forming <u>arrangements</u> of these letters where the first letter represents the president, the second the vice-president and the third the treasurer we could represent the solution to Problem 1 as follows:

> ABC, ABD, ABE, ACD, ACE, ADE, BCD, BCE, BDE, CDE
> ACB, ADB, AEB, ADC, AEC, AED, BDC, BEC, BED, CED
> CAB, DAB, BAE, DAC, EAC, EAD, DBC, EBC, EBD, ECD
> BAC, BAD, EAB, CAD, CAE, DAE, CBD, CBE, DBE, DCE
> BCA, BDA, EBA, CDA, CEA, DEA, CDB, CEB, DEB, DEC
> CBA, DBA, BEA, DCA, ECA, EDA, DCB, ECB, EDB, EDC.

Observe there are sixty different arrangements of the

letters. Another name for an arrangement is a <u>permutation</u>.
Hence we have 60 permutations of five different objects
taken 3 at a time. The answer to Problem 1 is that there are
60 ways to choose a president, vice-president, and a treasurer.

From the list of arrangements we can also answer the
question in Problem 2 of choosing a committee of three.
It is important to remember that within a committee there
is no order to positions. So that ABC is the same committee
as ACB, as CAB, as BAC, as BCA, as CBA. In fact all the
elements in Column one of our list under ABC represent the
same committee. Likewise for the second column and so on.
Hence there are only 10 distinct committees.

Actually we could have solved the two problems using the
fundamental principle of counting.

<u>PROBLEM 1.</u> Given 5 people there are 5 ways to choose a
president, then 4 ways to choose a vice-president and then
3 ways to choose a treasurer. Hence (5)(4)(3) = 60 ways to
choose all three positions.

<u>PROBLEM 2.</u> Once we have chosen a given three people from the
five there are six arrangements that we may make for the three
people. For example choose ABC. Now there are 3 choices
for the 1st position, 2 choices for the second and 1 choice
for the third, i.e., (3)(2)(1) = 6 arrangements of ABC. So
if we take the 60 arrangements of 5 people taken 3 at a time
and divide by 6, we have, 60/6 = 10 distinct committees of 3
people. We will refer to committees or subsets in which the
order <u>does not</u> matter as <u>combinations</u>. Thus the combinations
of 5 people taken 3 at a time is 10.

We now attempt to generalize the two problems.

<u>DEFINITION.</u> The symbol n! (read as "n" factorial) is defined
as follows:

i)   0! = 1

ii)   1! = 1

iii)   (k+1)! = k!(k+1) for every k ε Z$^+$

EXAMPLE: Evaluate 5!

Solution: 5! = 4!(5) = 3!(4)(5) = 2!(3)(4)(5)

= 1!(2)(3)(4)(5) = (1)(2)(3)(4)(5) = 120

EXAMPLE: Evaluate $\frac{9!}{6!}$

Solution: 9! = 6!(7)(8)(9). Hence $\frac{9!}{6!} = \frac{\not{6!}(7)(8)(9)}{\not{6!}} = 504$

Notice to solve Problem 1 in the case were there are n people with $\imath$ elected offices to fill we have

$$n(n-1)(n-2)\cdots(n-\imath+1) = \frac{n!}{(n-\imath)!}$$

ways of choosing candidates for the offices.

DEFINITION: Given a set with n different objects. The number of permutations (or arrangements) of n objects taken $\imath$ at a time is represented by $P(n,\imath)$. Read as permuations of n things taken $\imath$ at a time.

Note: To calculate $P(n,\imath)$ we use

$$\boxed{P(n,\imath) = \frac{n!}{(n-\imath)!} = n(n-1)(n-2)\cdots(n-\imath+1)}$$

To solve Problem 2 in the case where there are n people and we wish to choose committees (subsets) of $\imath$ people, we first consider the number of arrangements (permutations) of

208

n people taken n at a time, i.e., $P(n,n)$. Then we consider the number of ways a given subset of n people can be arranged, i.e., $P(n,n)$. Hence the number of committees (combinations) is

$$\frac{P(n,n)}{P(n,n)} = \frac{\frac{n!}{(n-n)!}}{\frac{n!}{0!}} = \frac{n!}{n!(n-n)!}$$

DEFINITION: Given a set with n different elements. The number of subsets (committees) containing n elements is represented by $C(n,n)$. Read as "combinations of n things taken n at a time."

Note: To calculate $C(n,n)$ we use

$$\boxed{C(n,n) = \frac{P(n,n)}{n!} = \frac{n!}{n!(n-n)!}}$$

EXAMPLE 1.   Evaluate

   (a)   P(6,3)
   (b)   C(6,3)

Solution: (a) $P(6,3) = \frac{6!}{(6-3)!} = \frac{6!}{3!} = \frac{(6)(5)(4)(3!)}{3!} = 120$

   (b) $C(6,3) = \frac{6!}{3!(6-3)!} = \frac{(6)(5)(4)(3!)}{(3)(2)(1)3!} = 20$

EXAMPLE 2.   Evaluate

   (a)   P(7,4)
   (b)   C(7,4)

Solution: (a) $P(7,4) = \frac{7!}{(7-4)!} = \frac{7!}{3!} = \frac{(7)(6)(5)(4)3!}{3!} = 840$

209

(b) $C(7,4) = \dfrac{7!}{4!3!} = \dfrac{(7)(6)(5)(4)3!}{(4)(3)(2)(1)3!} = 35$

EXAMPLE 3. How many 4-digit numbers can be formed using the digits 1, 2, 3, ..., 8, 9 without repetition?

Solution: This problem calls for an arrangement of 9 different objects taken 4 at a time. Hence

$$P(9,4) = \dfrac{9!}{5!} = \dfrac{(9)(8)(7)(6)5!}{5!} = 3024$$

EXAMPLE 4. How many ways can one obtain exactly 3 jacks in a 5 card hand?

Solution: To solve this problem we must think about it as two tasks: first to select 3 jacks from the 4 available; second to select the remaining 2 cards from the non-jacks, i.e., 48 cards.

TASK 1: Since the order of the jacks does not matter, we have

$$C(4,3) = \dfrac{4!}{3!1!} = \dfrac{(4)(3!)}{(3!)} = 4$$

TASK 2: Again since the order does not matter, we have

$$C(48,2) = \dfrac{48!}{2!46!} = \dfrac{(48)(47)46!}{(2)46!} = 1128$$

Then by the FPC we have

$$C(4,3) \cdot C(48,2) = (4)(1128) = 4512$$

ways of selecting 3 jacks in a 5 card hand.

EXAMPLE 5. How many 9 letter words can be formed using only the letters t, t, s, e, e, e, g, g, a?

Solution: If we were to make all the letters different by subscripting, i.e., t_1, t_2, s, e_1, e_2, e_3, g_1, g_2, a; then we are asking for the number of arrangements of 9 different objects taken 9 at a time which is

$$P(9,9) = \frac{9!}{0!} = 9!$$

However, $t_1 t_2 s e_1 e_2 e_3 g_1 g_2 a$ is indistinguishable from $t_2 t_1 s e_1 e_2 e_3 g_1 g_2 a$ when the subscripts are removed. Thus, once we have picked an arrangement there are 2! ways of rearranging the t's in which the arrangements are indistinguishable. Likewise, 3! ways for the e's and 2! ways for the g's. Thus the one arrangement $t_2 t_1 s e_1 e_2 e_3 g_1 g_2 a$ gives rise to $(2!)(3!)(2!)$ ways that will be exactly the same once the subscripts are removed. The same argument can be used for any subscripted arrangement that we start with. Thus, if there are x distinguishable arrangements (subscripts removed) there will be

$$(x)[(2!)(3!)(2!)] = 9!$$

arrangements when the subscripts are considered. Hence

$$x = \frac{9!}{2!3!2!} = 15120$$

distinguishable arrangments.

The solution can be generalized as follows: Suppose there are N objects of which N_1 are of one kind, N_2 of another kind, N_2 of a third kind and so on, where

$$N = N_1 + N_2 + N_3 + \ldots + N_t .$$

Then the number of distinguishable arrangements of the N objects is:

$$\frac{N!}{(N_1!)(N_2!)\ldots(N_t!)}$$

EXAMPLE 6. How many ways can 3 people be arranged at a circular table?

Solution:

are all the

same arrangement. So are

Thus, there are only two possible arrangements.

To generalize - we can calculate the number of arrangements in circular order by fixing the position of one object arbitrarily and then calculating the number of arrangements of the remaining objects as if they were in a straight line. Thus N distinct objects can be arranged in a circle in (N-1)! ways. However there are

$$\frac{(N-1)!}{2}$$

different arrangements of N keys on a keyring! Why?

EXAMPLE 7. How many committees of 5 can be formed from 10 men and 8 women if each committee is to have at most 3 men?

Solution: We may have committees with the following structures:

1. 0 men and 5 women

$$(_{10}C_0)(_{8}C_5) = \left(\frac{10!}{0!10!}\right)\left(\frac{8!}{5!3!}\right) = 56$$

2. 1 man and 4 women

$$(_{10}C_1)(_8C_4) = \frac{10!}{1!9!} \quad \frac{8!}{4!4!} = 700$$

3. 2 men and 3 women

$$(_{10}C_2)(_8C_3) = \frac{10!}{2!8!} \quad \frac{8!}{3!5!} = 2520$$

4. 3 men and 2 women

$$(_{10}C_3)(_8C_2) = \frac{10!}{3!7!} \quad \frac{8!}{2!6!} = 3360$$

Hence, the total number of committees is 56 + 700 + 2520 + 3360 = 6636

SELF-TEST:

1. Evaluate

     (a) P(12,8)   (b) C(12,8)

                                            ANS: (a) 19958400
                                                  (b) 495

2. How many ways can 10 people be seated at a circular table?

                                            ANS: 362880

3. How many committees of five can be formed from 5 men
    and 9 women if a given committee must contain <u>exactly</u>
    2 men?

                                    ANS: $C(5,2) \cdot C(9,3) = 840$

213

EXERCISES 4.2

1. Evaluate:

    (a) P(8,4)   (b) C(8,4)

2. Evaluate:

    (a) P(10,7)   (b) C(10,7)

3. Evaluate:

    (a) P(9,3)   (b) C(9,3)

4. Evaluate:

    (a) P(7,5)   (b) C(7,5)

5. How many 3-digit numbers can be formed using the digits 1, 2, 3, 4, 5, 6, 7, 8, 9 without repetition?

6. How many 5-digit numbers can be formed using the digits 1, 2, 3, 4, 5, 6, 7, 8, 9 without repetition?

7. How many different arrangements can be formed on a shelf with space for 4 books if there are 7 different books available?

8. How many different arrangements can be formed on a shelf with space for 5 books if there are 8 different books available?

9. How many ways can a president, vice president and secretary be chosen from 10 candidates, if no candidate may hold two offices simultaneously?

10. In how many ways can a president, vice president and secretary be chosen from 12 people, if no one may hold a dual position?

11. How many permutations can be formed from the letters of the word committee?

12. How many permutations can be formed from the letters in the word Mississippi?

13. How many ways can 7 people be arranged at a circular table?

14. How many ways can 9 people be arranged at a circular table?

15. How many ways can a committee of 3 be chosen from a set of 12 people?

16. How many ways can a committee of 5 be chosen from a set of 10 people?

17. From a group of 10 boys and 8 girls, how many different committees of 7 can be chosen which contain exactly 4 boys.

18. From a group of 10 boys and 8 girls how many different committees of 7 can be chosen with exactly 3 girls?

19. From a group of 10 boys and 8 girls how many different committees of 7 can be chosen with exactly 2 boys and 2 girls?

20. From a group of 10 boys and 8 girls how many different committees of 7 can be chosen with exactly 3 boys and 3 girls?

21. How many subsets does a set with 4 elements have?

22. How many subsets does a set with 6 elements have?

23. How many 5-card hands can be dealt from a deck of 52 cards?

24. How many ways can one obtain exactly 3 kings in a 5-card hand?

25. How many ways can one obtain exactly 3 aces in a 5-card hand?

26. From an urn containing 8 black and 5 white balls, in how many ways can we draw a set of 6 balls of which 3 are white and 3 are black?

27. From an urn containing 8 black and 5 white balls, in how many ways can we draw a set of 5 balls of which 2 are black and 3 are white?

28. How many committees of four can be formed from 12 men and 6 women if each committee is to have at most 3 men?

29. How many committees of four can be formed from 12 men and 6 women if each committee is to have at most 3 women?

30. How many straight lines are determined by the vertices of a regular pentagon?

4.3 PROBABILITY MEASURES AND FINITE SAMPLE SPACES

In order to discuss the proability of a given experiment, either real or conceptual, we need to define a structure which will be called a sample space.

DEFINITION. A <u>sample space S</u> associated with a given experiment is a set such that

    i)   each element of S corresponds to a possible outcome of the experiment and

    ii) each possible outcome of the experiment corresponds to one and only one element of S.

DEFINITION. An <u>event</u> is a subset of a sample space.

It is possible for a given experiment to be associated with many different sample spaces that meet the requirements of the above definition.

EXAMPLE 1. Suppose two "fair" dice are rolled. Construct a sample space.

Solution. One possible sample space could correspond to the sum appearing on the dice, i.e., $S_1 = \{2,3,4,5,6,7,8,9,10,11,12\}$. Another possible sample space might be constructed as follows: suppose we color one die red and the other green; then we record the number of the red die first and the number on the green die second. We could then represent the sample space as a set of ordered pairs

$$S_2 = \left\{ \begin{array}{cccccc} (1,1), & (1,2), & (1,3), & (1,4), & (1,5), & (1,6) \\ (2,1), & (2,2), & (2,3), & (2,4), & (2,5), & (2,6) \\ (3,1), & (3,2), & (3,3), & (3,4), & (3,5), & (3,6) \\ (4,1), & (4,2), & (4,3), & (4,4), & (4,5), & (4,6) \\ (5,1), & (5,2), & (5,3), & (5,4), & (5,5), & (5,6) \\ (6,1), & (6,2), & (6,3), & (6,4), & (6,5), & (6,6) \end{array} \right\}$$

Notice that either S_1 or S_2 is a legitimate sample space for the given experiment.

EXAMPLE 2. Flip a dime and a nickel. Construct a sample space for this experiment.

Solution. One sample space might record the number of heads obtained, i.e., $S_3 = \{0,1,2\}$. Another possible space might record first what appears on the nickel and second what appears on the dime. Hence the sample space would be a set of ordered pairs, i.e., $S_4 = \{HH,HT,TH,TT\}$. Again both S_3 and S_4 are legitimate sample spaces.

DEFINITION. Suppose S is a sample space. Then a <u>simple event</u> of S is a subset of S containing <u>exactly one element</u>.

Once we have constructed a sample space for a given experiment, we then go about weighting each simple event in a way that will tell something of its likelihood of occurrence.

DEFINITION. Let S be a sample space and let W be a simple event in S. To each W in S we assign a real number called the <u>probability of W</u> symbolized by P(W) such that

   i) $P(W) \geq 0$

  ii)   The sum of all the probabilities assigned to all simple events of S is 1, i.e., if

$$S = W_1 \cup W_2 \cup \ldots \cup W_k$$

where W_i's are simple events then

$$P(W_1) + P(W_2) + \ldots + P(W_k) = 1.$$

How the probability of a simple event is assigned is a

completely arbitrary matter as long as the assignment meets the above two conditions.

If we were dealing with a real experiment we might assign P(W) by using the notion of the "relative frequency" of the event. If we have a conceptual experiment then we usually try first to construct a sample space in which each simple event has an equally likely weight of happening. Thus, if the sample space contains n elements and W is a simple event, then

$$P(W) = \frac{1}{n}.$$

DEFINITION. Let sample space $S = W_1 \cup W_2 \cup \ldots W_n$ where W_i's are simple events. Then the probability of an event $E \subseteq S$ is given by

$$P(E) = \sum_{W_i \subseteq E} P(W_i)$$

i.e., the probability of event E is the sum of the probabilities of the simple events that make up E.

DEFINITION. Since ϕ is an event of any sample space we define $P(\phi) = 0$.

EXAMPLE 3. Roll a fair die. What is the probability of obtaining a 3.

Solution. Note that there could be an infinite number of answers to this question depending on the sample space chosen as well as the probability measure one assigns to the elements of the given sample space.. However, in most experiments there is what one might call a "natural sample space" and this is the one with which we will usually work. Hence, let

$$S = \{1,2,3,4,5,6\}.$$

Since each simple event in S has an equal weight of happening in our experiment we assign to a simple event W, $P(W) = \frac{1}{6}$. Thus, $P(\{3\}) = \frac{1}{6}$.

EXAMPLE 4. Suppose two dice are rolled. Find the probability that the sum on the dice is 6.

Solution. Return to Example 1. We choose sample space S_2 rather than S_1 because in S_1 it is harder to decide the weight of a simple event. S_2 gives us a better picture of the way sums actually occur. S will be the natural sample space. Hence P({sum of six}) =

$$P(\{(5,1),(4,2)(3,3),(2,4),(1,5)\}) = \frac{5}{36}$$

NOTE: If we are dealing with a sample space containing N simple events and each simple event is assigned the same probability, then the probability of an event E which contains k elements is given by

$$P(E) = \frac{k}{n} = \frac{n(E)}{n(S)}$$

The method noted above is very useful, especially in a situation in which the sample space is too large to write out.

EXAMPLE 5. Suppose 5 coins are tossed together. What is the probability of obtaining exactly two heads?

Solution. First we color the coins in the colors red, white, blue, green, and yellow. Then we record the result of each coin in a 5-tuple in the order red, white, blue, green and yellow. Since there are five positions in the 5-tuple and each with a possibility of being either a H or a T by the FPC there are $(2)(2)(2)(2)(2) = 2^5 = 32$ elements in the sample space. Then to each simple event in the sample we assign $P(W) = \frac{1}{32}$. Now if E is the event exactly two heads, how many simple events are there in E? Well, out of 5 positions we must choose exactly two to be heads hence $C(5,2) = 10$ ways to have exactly 2 heads. In actuality it would not have been hard to count directly the elements in event E since

$$E = \{HHTTT, HTTTH, HTHTT, HTTHT, THHTT, TTHHT, TTTHH,$$
$$THTHT, TTHTH, THTTH\}$$

Thus $P(E) = \frac{n(E)}{n(S)} = \frac{10}{32}$ or $P(E) = C(5,2)\frac{1}{32} = \frac{10}{32}$

DEFINITION. If two events, A and B, can not occur together we call A and B <u>mutually exclusive</u> events. If A and B are mutually exclusive then $A \cap B = \emptyset$.

THEOREM 1. IF A and B are mutually exclusive events in a sample space S, where each simple event has the same probability, then

$$P(A \cup B) = P(A) + P(B).$$

Proof: $P(A \cup B) = \frac{n(A \cup B)}{n(S)} = \frac{n(A)+n(B)}{n(S)}$

$$= \frac{n(A)}{n(S)} + \frac{n(B)}{n(S)}$$

$$= P(A) + P(B)$$

THEOREM 2. If A and B are two events in a sample space S, where each simple event has been assigned the same probability, then

$$P(A \cup B) = P(A) + P(B) - P(A \cap B)$$

Proof: Left as an exercise.

THEOREM 3. Let E be an event in a sample space S, where each simple event has been assigned the same probability. Let $\overline{E}$ represent the complement of E in S. Then

$$P(E) = 1 - P(\overline{E})$$

Proof: $P(S) = \dfrac{n(S)}{n(S)} = 1$

But $E \cup \overline{E} = S$ and $E \cap \overline{E} = \emptyset$

Hence $P(S) = P(E \cup \overline{E}) = P(E) + P(\overline{E})$

i.e., $1 = P(E) + P(\overline{E})$ or $1 - P(\overline{E}) = P(E)$.

It should be noted that the above theorems actually hold in any sample space with a legitimate probability measure, although the proofs would then be handled differently.

EXAMPLE 6. Suppose 5 coins are tossed. Find the probability that at least 1 coin falls heads.

Solution. Let S be the sample space from Example 5. Let E be the event that at least one coin falls heads. Then $\overline{E}$ is the event no coin falls head, i.e.g all coins are tails. Hence $\overline{E} = \{(,T,T,T,T,T)\}$, and so $P(\overline{E}) = \dfrac{1}{32}$. Thus,

$$P(E) = P - P(\overline{E}) = 1 - \frac{1}{32} = \frac{31}{32}.$$

DEFINITION. Let E be any event in a sample space S. Let $\overline{E}$ be the complement of E. Then the odds for E are $\dfrac{P(E)}{P(\overline{E})}$. Also the odds against E are $\dfrac{P(\overline{E})}{P(E)}$.

EXAMPLE 7. If five cars are dealt from an ordinary deck, find the odds that four kings are in the hand.

Solution. Consider a sample space S consisting of all the

222

possible 5 card hands. Then n(S) = C(52,5). To find the number of hands having four kings we note that there is C(4,4) = 1 way of choosing the kings and C(48,1) = 48 ways of choosing the fifth card. Hence there are 48 5-card hands with 4 kings. Let E = event that a hand contains 4 kings.

$$P(E) = \frac{C(4,4)C(48,1)}{C(52,5)} = \frac{48}{2598960}$$

$$P(\bar{E}) = 1 - \frac{48}{2598960} = \frac{2598912}{2598960}$$

Hence, the odds for E = $\frac{48}{2598912}$ = .00001847 or about $\frac{18}{1,000,000}$

SELF-TEST:

1. In rolling a pair of dice what is the probability of a total of five or less?

    ANS: $\frac{5}{18}$

2. If 5 cards are drawn from an ordinary deck, find the probability all 5 cards are hearts.

    ANS: $\frac{1287}{2598960}$

3. If 3 coins are tossed, find the odds that all three turn up heads.

    ANS: $\frac{1}{7}$

EXERCISES 4.3

1. Find the probability of rolling a five with a single die.

2. In rolling a single die what is the probability of an even number?

3. In rolling a pair of dice what is the probability of a total of 8 or more?

4. In rolling a pair of dice what is the probability of a total of 4 or less?

5. In the toss of two fair coins what is the probability of exactly two heads?

6. Find the probability of obtaining exactly one head in the toss of two coins?

7. If 5 cards are drawn from an ordinary deck of 52 playing cards, what is the probability that all 5 cards are spades?

8. If 5 cards are drawn from an ordinary deck of 52 playing cards what is the probability that all 5 cards are the same suit?

9. If 5 cards are drawn from an ordinary deck of 52 playing cards what is the probability that the cards will contain exactly three aces?

10. If 5 cards are drawn from an ordinary deck of 52 playing cards what is the probability that the cards will form a royal straight flush?

11. If 5 cards are drawn from an ordinary deck of 52 playing cards what is the probability that the cards will

contain a pair or better?

12. If 3 fair coins are tossed what is the probability of exactly 2 heads being obtained?

13. If 3 fair coins are tossed what is the probability of at least 2 heads being obtained?

14. If 10 coins are tossed what is the probability of obtaining exactly zero heads?

15. If 6 people are seated at random at a round table, what is the probability that a certain 2 will be neighbors?

16. If a penny, a dime, and a quarter are tossed find the probability of at least two heads.

17. If a penny, a dime and a quarter are tossed find the probability of at most two heads.

18. If 10 people are seated at a round table, what is the probability that a certain 2 will be neighbors?

19. Six different books are to be placed at random on a bookshelf with space for six books. What is the probability that a certain 2 books will be next to each other?

20. If eight different books are to be placed at random on a bookshelf with space for eight books what is the probability that a certain two books will be next to each other?

21. If four balls are drawn simultaneously from an urn containing 5 red and 4 white balls, what is the probability that exactly 3 will be red?

22. If four balls are drawn simultaneously from an urn

containing 5 red and 4 white balls what is the probability that at least three will be red?

23. If four balls are drawn simultaneously from an urn containing 5 red and 4 white balls, what is the probability that at most three will be red?

24. Five letters together with corresponding envelopes are addressed to five different people. Suppose the letters are dropped onto the floor and then as they are picked up randomly inserted into an envelope. Find the probability that

   (a)   every letter is inserted into the right envelope
   (b)   that exactly one letter is put in the right envelope.

25. Find the odds for rolling a sum of five with a pair of dice.

26. Suppose 4 machines are chosen at random from a shipment of 20 machines, 6 of which are defective. What are the odds that all four machines will be defective?

27. Suppose the odds against E are 2 to 1. Find the P(E).

28. If five cards are dealt from an ordinary deck, find the odds that a flush is dealt.

29. If five cards are dealt from an ordinary deck, find the odds that the hand contains 4 of the same kind.

30. A gambler offers 1 to 3 odds that event A will occur and 7 to 5 odds that event B will occur. If A and B are mutually exclusive events, what odds should be given that A or B will occur?

31. A gambler offers 2 to 5 odds that event A will occur and 8 to 5 odds that event B will occur. If A and B are mutually exclusive events, what odds should be given that A or B will occur?

32. Show that if the odds for event E are a/b then the

$$P(E) = \frac{a}{a+b}.$$

4.4 INDEPENDENT EVENTS AND CONDITIONAL PROBABILITY

OBJECTIVES:

1. To compute conditional probabilities.
2. To define independent events.

Consider the following problem:

PROBLEM. Suppose a fair coin is tossed 4 times. Find the probability of getting at least 3 heads given that the first toss is a head.

Solution. Let A be the event of getting at least 3 heads. Then

$$A = \{HHHT, HHTH, HTHH, THHH, HHHH\}$$

Let B = event that the first toss is heads.

$$B = \{HHHT, HHTH, HTHH, HHHH, HHTT, HTHT, HTTH, HTTT\}$$

Now let us look at the sample space for the experiment together with events A and B in a Venn diagram.

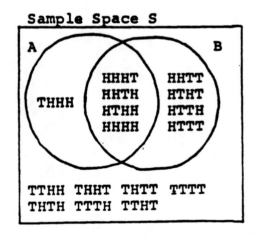

Sample Space S

*NOTE: Read P(A|B) "probability of A given B has occurred."

Notice that there are 16 elements in the same space S. We wish to compute P(A|B), i.e., the probability of event A, given that event B has occurred. Once we have been given that event B has occurred we can restrict our attention to only those elements in S that are in event B. In this way B becomes the "reduced sample space". Now to find P(A|B) we look at where A intersects B, i.e., there are 4 possibilities out of the 8 possibilities in the "reduced sample space" B. So

$$P(A|B) = \frac{4}{8} = \frac{n(A \cap B)}{n(B)}.$$

Since we are dealing with a sample space in which all simple events have an equally likely weight of happening:

$$\frac{n(A \cap B)}{n(S)} = P(A \cap B) \quad \text{and} \quad \frac{n(B)}{n(S)} = P(B)$$

Thus,

$$P(A|B) = \frac{\dfrac{n(A \cap B)}{n(S)}}{\dfrac{n(B)}{n(S)}} = \frac{P(A \cap B)}{P(B)}$$

DEFINITION. Let A and B be events of a sample space S. Suppose P(B) ≠ 0. Then the conditional probability of A given B is computed by

$$P(A|B) = \frac{P(A \cap B)}{P(B)}$$

DEFINITION. Two non-empty events will be called independent if the occurrence or non-occurrence of the first has no effect on the occurrence or non-occurrence of the second, i.e., A and B are independent events if P(A|B) = P(A) or P(B|A) = P(B).

THEOREM. A and B are independent events if and only if P(A∩B) = P(A)P(B).

Proof. If A and B are independent then $P(A|B) = P(A)$.

But $P(A|B) = \dfrac{P(A \cap B)}{P(B)}$ so $P(A) = \dfrac{P(A \cap B)}{P(B)}$ and

$P(A)P(B) = P(A \cap B)$.

On the other hand if $P(A \cap B) = P(A)P(B)$ then $P(A|B) =$
$\dfrac{P(A \cap B)}{P(B)} = \dfrac{P(A)P(B)}{P(B)} = P(A)$ Hence A and B are independent.

EXAMPLE 1. A green die and a blue die are rolled. Find the probability of obtaining a sum greater than 10 given that the blue die shows a 5.

Solution: Let the sample space

$$S = \left\{ \begin{array}{l} (1,1),\ldots(1,6) \\ \cdot \\ \cdot \\ \cdot \\ \cdot \\ (6,1),\ldots(6,6) \end{array} \right\}$$

where the first element is the green die and the second is the blue die. Let A be event sum greater than 10 = {(6,6),(6,5),(5,6)}. Let B be event blue die shows a 5 = {(1,5),(2,5),(3,5),(4,5),(5,5), (6,5)}. Then $A \cap B$ = {(6,5)}. Hence

$$P(A|B) = \frac{P(A \cap B)}{P(B)} = \frac{\frac{1}{36}}{\frac{6}{36}} = \frac{1}{6}$$

EXAMPLE 2. A dime is tossed three times. Find the probability that the third toss is heads given that the first two are heads.

Solution. Let the sample space S = {HHH,THH,HTH,HHT,HTT,THT,TTH,TTT} Let A be the event that the third toss is heads, i.e., A = {HHH,THH,HTH,TTH}. Let B be the event that the first two tosses are heads, i.e., B = {HHH,HHT}. Then $A \cap B$ = {HHH}. Then

$$P(A|B) = \frac{P(A \cap B)}{P(B)} = \frac{\frac{1}{8}}{\frac{2}{8}} = \frac{1}{2}$$

EXAMPLE 3.   Decide whether A and B in Example 2 are independent events.

Solution. $P(A) = \frac{4}{8} = \frac{1}{2}$

$\qquad P(A \cap B) = \frac{1}{8}$

$\qquad P(B) = \frac{2}{8} = \frac{1}{4}$

Check: $P(A \cap B) \stackrel{?}{=} P(A)P(B)$

$$\frac{1}{8} \stackrel{?}{=} \left(\frac{1}{2}\right)\left(\frac{1}{4}\right)$$

YES!

So A and B are independent events. This result should be somewhat intuitively obvious from the nature of the experiment.

EXAMPLE 4.   Suppose two cards are drawn at random from an ordinary deck of cards. If one of them is a king, what is the probability that both of them are kings? We offer the following solutions to this problem.

Solution #1. Let A be the event "both cards are kings". Let B be the event "at least one card is a king". Then A ∩ B = A. We wish to find P(A|B). To find P(A) consider there are C(52,2) draws of two cards and C(4,2) draws of 2 kings. Hence

$$P(A) = \frac{C(4,2)}{C(52,2)} = \frac{1}{221}$$

To find P(B) consider $\overline{B}$, i.e., the event neither card is a king.

$$P(\bar{B}) = \frac{C(48,2)}{C(52,2)} = \frac{188}{221}$$

So that $P(B) = 1 - P(\bar{B}) = 1 - \frac{188}{221} = \frac{33}{221}$.

Thus $P(A|B) = \frac{P(A \cap B)}{P(B)} = \frac{P(A)}{P(B)} = \frac{\frac{1}{221}}{\frac{33}{221}} = \frac{1}{33}$

Solution #2. This solution uses the formula $P(D|C) = \frac{P(C \cap D)}{P(C)}$

in the form $P(C \cap D) = P(C)P(D|C)$. Also we use the idea of a tree diagram which breaks the problem down into mutually exclusive events. Let C be 1st card a king; then $\bar{C}$ is 1st card non-king. Let D be 2nd card a king; then $\bar{D}$ is 2nd card non-king. Our tree diagram is as follows:

$$C \cap D \qquad P(C \cap D) = P(C)P(D|C) = \frac{4}{52}\;\frac{3}{51}$$
$$C \cap \bar{D} \qquad P(C \cap \bar{D}) = P(C)P(\bar{D}|C) = \frac{4}{52}\;\frac{48}{51}$$
$$\bar{C} \cap D \qquad P(\bar{C} \cap D) = P(\bar{C})P(D|\bar{C}) = \frac{48}{52}\;\frac{4}{51}$$
$$\bar{C} \cap \bar{D} \qquad P(\bar{C} \cap \bar{D}) = P(\bar{C})P(\bar{D}|\bar{C}) = \frac{48}{52}\;\frac{47}{51}$$

Note if we write the probabilities of the events on their respective branches, then the probability of following a particular path is the product of the probabilities on the branches.

Now if A and B are defined as in Solution 1, then

$$P(A|B) = \frac{P(A \cap B)}{P(B)} = \frac{P(A)}{P(B)}$$

$$P(A) = P(C \cap D) = \frac{4}{52}\;\frac{3}{51}$$

$$P(B) = P(C \cap D) + P(C \cap \bar{D}) + P(\bar{C} \cap D) = \frac{(4)(3)+(48)(4)+4(48)}{(52)(51)}$$

Hence $P(A|B) = \dfrac{12}{12+(48)(4)+(48)(4)} = \dfrac{1}{1+16+16} = \dfrac{1}{33}$

EXAMPLE 5. Suppose three bags contain colored "jaw-breakers" as follows:

Bag I: 3 red, 2 green, 4 yellow
Bag II: 2 red, 5 green, 3 yellow
Bag III: 5 red, 4 green, 2 yellow

A bag is chosen at random and then a "jaw-breaker" is selected. If the "jaw-breaker" selected is green, find the probability that it came from Bag III.

Solution. We use a tree diagram to find the probability of Bag III given a green "jaw-breaker", i.e., P(III|G).

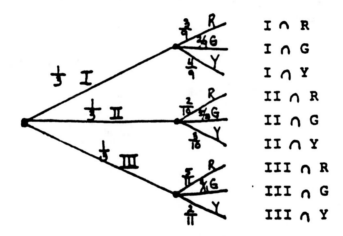

$$P(III|G) = \frac{P(III\cap G)}{P(G)} = \frac{P(III\cap G)}{P(I\cap G)+P(II\cap G)+P(III\cap G)}$$

$$= \frac{\left(\frac{1}{3}\right)\left(\frac{4}{11}\right)}{\left(\frac{1}{3}\right)\left(\frac{2}{9}\right)+\left(\frac{1}{3}\right)\left(\frac{5}{10}\right)+\left(\frac{1}{3}\right)\left(\frac{4}{11}\right)}$$

$$= \frac{(4)(90)}{(2)(10)(11)+5(9)(11)+4(9)(10)}$$

$$= \frac{360}{1075} = \frac{72}{215}$$

SELF-TEST:

1. Urn I contains 5 red balls and 3 green balls. Urn II contains 4 red and 2 green. Suppose an urn is chosen at random and then a ball selected. Find the probability that Urn II was chosen given a green ball was selected.

$$\text{ANS:} \quad \frac{8}{17}$$

2. Assume that each birth to the same parents is an independent event. Also assume $P(\text{Boy}) = P(\text{Girl}) = \frac{1}{2}$. Suppose a family has two children.

   (a) Find the probability that both are girls given one is a girl.

   (b) Find the probability that both are girls given the younger is a girl.

$$\text{ANS:} \quad \text{(a)} \quad \frac{1}{3}$$

$$\text{(b)} \quad \frac{1}{2}$$

EXERCISES 4.4

1. A blue die and a red die are tossed together. Find the probability that the sum is greater than 7, given that the blue die shows a 5.

2. A blue die and a red die are tossed together. Find the probability that the sum is less than 7, given that the blue die shows a 3.

3. A blue die and a red die are tossed together. Find the probability that the sum is 10, given that the blue die is even.

4. A blue die and a red die are tossed together. Find the probability that the sum is 8, given that the blue die is even.

5. Two fair coins, a dime and a nickel, are tossed. Find the probability that both coins are heads given that the dime shows heads.

6. Two fair coins, a dime and a nickel are tossed. Find the probability that both coins show heads if it is given that one shows heads.

7. Suppose two cards are drawn at random from an ordinary deck of cards. Find the probability that both are kings, given that both cards are face cards (i.e., jack, queen, king).

8. Suppose two cards are drawn at random from an ordinary deck of cards. Find the probability that both cards are hearts, given that both cards are red.

9. Suppose two cards are drawn at random from an ordinary

235

deck of cards. Find the probability that both cards are
hearts, given that both cards are face cards.

10. Suppose two cards are drawn at random from an ordinary
    deck of cards. Find the probability that both cards
    are not face cards, given that both cards are hearts.

11. Suppose two cards are drawn at random from an ordinary
    deck of cards. Find the probability that both cards
    are face cards, given that both cards are red.

12. A fair coin is tossed three times. Find the probability
    of getting at least two tails, given that the first toss
    is tails.

13. A fair coin is tossed four times. Find the probability
    of getting at least three heads, given that the first
    toss is heads.

14. Suppose that 75% of the voters in a certain city voted
    for the Republican Mayor. In the same election suppose
    that 40% voted for the Republican Mayor and for the sales
    tax issue. Find the probability that a voter selected
    at random voted for the sales tax given that he voted
    for the Republican Mayor.

15. A fair coin is tossed three times. Find the probability
    that the third toss is tails, given that the first two
    tosses are tails.

16. An urn contains four white balls and two green ones.
    A set of 3 balls is drawn consecutively without re-
    placement. Find the probability that the third ball
    is white given that the first two are white.

17. An urn contains 3 white and 5 green balls. A set of

236

four balls is drawn consecutively without replacement. Find the probability that the fourth ball is green, given that the first three are green.

18. Two defective light bulbs get put in with 5 good light bulbs. The bulbs are tested one by one until all the defective ones are located. What is the probability that the sixth bulb tested is the second defective one?

19. Cards are dealt one at a time from an ordinary deck. Find the probability that the fourth card dealt is the first heart.

20. Cards are dealt one at a time from an ordinary deck. Find the probability that the sixth card dealt is the first king.

21. Suppose that 3 defective light bulbs are mixed with 4 good ones. The bulbs are tested one by one until one finds the two defectives. Draw a tree diagram for this experiment. Find the probability that it takes four tests to find the defectives.

22. Let A and B be independent events such that $P(A) = .6$ and $P(b) = .5$. Find $P(A \cap B)$.

23. A baseball player calculates his chances at bat as follows: probability of getting a hit is .30 of getting a walk is .12, and of being out is .58 Suppose the player is at bat four times and each time is considered an independent event. Find the probability that the player gets one hit, one walk, and is out twice.

24. Show that if A and B are independent events, so are A and $\overline{B}$, $\overline{A}$ and B, $\overline{A}$ and $\overline{B}$.

25. In a study of 2000 students in a Lousiana State University, students were classified as from Lousiana, outside Louisiana, those who had studied French in high school and those who had not. The data is shown as follows:

	French in high school (F)	No French in high school ($\overline{F}$)
From La. (L)	1834	68
Not From LA ($\overline{L}$)	18	80

(a) Find $P(L|F)$
(b) $P(F)$
(c) $P(L)$
(d) Are L and F independent events?

26. An urn contains 8 blue balls and 4 red balls. Four balls are drawn one after the other. Find the probability that the first three are red and the fourth is blue.

27. Urn I contains 8 balls numbered 1 through 8. Urn II contains 5 balls numbered 1 through 5. First an urn is selected at random and then a ball is chosen. Suppose the number on the ball is 5. What is the probability that it came from Urn II?

28. Suppose three bags contain colored marbles as follows:

    Bag I: 2 red, 3 green, 3 yellow
    Bag II: 1 red, 4 green, 5 yellow
    Bag III: 6 red, 1 green, 2 yellow

If a yellow marble is selected, what is the probability that it came from Bag III?

29. Assume that each birth to the same parents is an independent event with P(boy) = P(girl) = .5. Suppose a

238

family has three children. Find the probability that
all three are boys given that at least one is a boy.

4.5 Random Variables

OBJECTIVES:

1. *Define random variables.*
2. *Define mathematical expectation.*
3. *Define variance and standard deviation.*

DEFINITION. Let S be a sample space associated with a given experiment. Let X be a function assigning to every $s \in S$ a real number $X(s)$. Then X is called a random variable.

EXAMPLE 1. Suppose four coins are tossed together. Construct a random variable X which indicates the number of heads showing when the coins are tossed.

Solution. Let X be the random variable defined as follows:

X(HHHH) = 4	X(HHTT) = 2	X(HTTT) = 1
X(HHTH) = 3	X(HTHT) = 2	X(THTT) = 1
X(HHHT) = 3	X(THHT) = 2	X(TTHT) = 1
X(HTHH) = 3	X(THTH) = 2	X(TTTH) = 1
X(THHH) = 3	X(TTHH) = 2	X(TTTT) = 0
	X(HTTH) = 2	

DEFINITION. P(X=t) read as "the probability that the random variable X assumes a value of t" is given by

$$P(X=t) = P(X^{-1}\{t\}).$$

EXAMPLE 2. Using the random variable X from Example 1, find P(X=3).

Solution. $P(X=3) = P(X^{-1}\{3\})$. But $X^{-1}\{3\} = \{HHHT, HHTH, HTHH, THHH\}$. Hence $P(X=3) = \frac{4}{16} = \frac{1}{4}.$

EXAMPLE 3. Using the random variable X from Example 1, find P(X=10).

Solution. $P(X=10) = P(X^{-1}\{10\}) = P(\emptyset) = 0$, i.e., since X never assumes the value of 10 then $P(X=10) = 0$.

DEFINITION. Let X be a random variable. The probability function P_X for the random variable X is given by

$$P_X(t) = P(X=t)$$

for any real number t.

DEFINITION. Let X be a random variable with a range $\{t_1, t_2, t_3, \ldots, t_n\}$. Let P_X be the probability function for X. Then the mathematical expectation (expected value), E(X), for the random variable X is given by

$$E(X) = \sum_{i=1}^{n_1} t_i P_X(t_i)$$

or $\quad E(X) = t_1 P_X(t_1) + t_2 P_X(t_2) + \ldots + t_n P_X(t_n).$

The expected value can be thought of as the weighted average of the possible outcomes of a random variable.

EXAMPLE 4. Suppose a pair of dice are rolled. What is the expected value of the random variable which assigns to each element in the sample space the sum of the dice?

Solution. The range of the random variable X is $\{2,3,4,5,6,7,8,9,10,11,12\}$

$$P_X(2) = P(X=2) = \frac{1}{36} \qquad P_X(8) = P(X=8) = \frac{5}{36}$$

$$P_X(3) = P(X=3) = \frac{2}{36} \qquad P_X(9) = P(X=9) = \frac{4}{36}$$

241

$$P_X(4) = P(X=4) = \frac{3}{36} \qquad\qquad P_X(10) = P(X=10) = \frac{3}{36}$$

$$P_X(5) = P(X=5) = \frac{4}{36} \qquad\qquad P_X(11) = P(X=11) = \frac{2}{36}$$

$$P_X(6) = P(X=6) = \frac{5}{36} \qquad\qquad P_X(12) = P(X=12) = \frac{1}{36}$$

$$P_X(7) = P(X=7) = \frac{6}{36}$$

Thus, $E(X) = 2\left(\frac{1}{36}\right) + 3\left(\frac{2}{36}\right) + 4\left(\frac{3}{30}\right) + 5\left(\frac{4}{36}\right) + 6\left(\frac{5}{36}\right) + 7\left(\frac{6}{36}\right) + 8\left(\frac{5}{36}\right)$
$$9\left(\frac{4}{36}\right) + 10\left(\frac{3}{36}\right) + 11\left(\frac{2}{36}\right) + 12\left(\frac{1}{36}\right) = \frac{252}{36} = 7$$

Observe that the E(X) of X is a weighted average of the numbers$\{2,3,\ldots,12\}$ where 2 occurs 1 time, 3 occurs 2 times, 4 occurs 3 times, 5 occurs 4 times, etc. Expected value is also called the mean value of the random variable.

EXAMPLE 5. Suppose a die is rolled. What is the expected value of the random variable which assigns the number on the die to the elements in the sample space?

Solution. The range of the random variable X = {1,2,3,4,5,6}.

$$P_X(1) = P(X=1) = \frac{1}{6}$$

$$P_X(2) = P(X=2) = \frac{1}{6}$$

$$P_X(3) = P(X=3) = \frac{1}{6}$$

$$P_X(4) = P(X=4) = \frac{1}{6}$$

$$P_X(5) = P(X=5) = \frac{1}{6}$$

$$P_X(6) = P(X=6) = \frac{1}{6}$$

So $E(X) = 1\left(\frac{1}{6}\right) + 2\left(\frac{1}{6}\right) + 3\left(\frac{1}{6}\right) + 4\left(\frac{1}{6}\right) + 5\left(\frac{1}{6}\right) + 6\left(\frac{1}{6}\right) = \frac{21}{6} = 3.5.$
Note that E(X) is not necessarily in the range of the random variable.

EXAMPLE 6. Suppose on the roll of a single die a gambler agrees to pay $10 if a 3 is rolled. If it costs $2 to play this game find the expected value of the net winnings

Solution. Note that net winnings = payoff - cost to play. Hence net winnings = $10 - $2 = $8 or 0 - $2 = -$2. Let X be a random variable such that

$$X(\delta) = \begin{cases} 8 \text{ if } \delta = 3 \\ -2 \text{ if } \delta \neq 3, \text{ i.e., } \delta = 1,2,4,5,6. \end{cases}$$

$$E(X) = 8\left(\frac{1}{6}\right) + (-2)\left(\frac{5}{6}\right) = \frac{-2}{6} = -\$.33\frac{1}{3}$$

Thus the expected loss per game on the long-run average is $.33\frac{1}{3}$. So on a run of 1000 games one would expect to be <u>in the hole</u> $333.33.

<u>DEFINITION.</u> If X is a random variable representing the net winnings in a game, then the game is considered fair if and only if E(X) = 0.

EXAMPLE 7. The game in Example 6 is not a fair game. To determine the price C to make it a fair game we solve the equation

$$E(X) = (10-c)(\frac{1}{6}) + (-c)(\frac{5}{6}) = 0$$
$$10-c-5c = 0$$
$$10-6c = 0$$
$$c = \frac{10}{6}$$
$$= \$1.67$$

i.e., $1.60 would be a fair price to pay to play the game.

<u>DEFINITION.</u> Let X be a random variable such that $E(X) = \bar{X}$. Let the range of X be $\{t_1, t_2, \ldots, t_n\}$. Then the variance of X, Var(x), is given by

$$Var(X) = \sum_{i=1}^{n_1} (t_i-\bar{X})^2 P_X(t_i)$$

i.e., $\text{Var}(X) = (t_1 - \bar{X})^2 P_X(t_1) + \ldots + (t_n - \bar{X})^2 P_X(t_n)$

DEFINITION. Let X be a random variable as above. Then the standard deviation of X, SD(X), is given by

$$SD(X) = \sqrt{\text{VAR}(X)}$$

EXAMPLE 8. Find the Var(X) and the SD(X) for the random variable X used in Example 4.

Solution. From Example 4 we know $E(X) = \bar{X} = 7$. So

$$\text{Var}(X) = (2-7)^2\left(\frac{1}{36}\right) + (3-7)^2\left(\frac{2}{36}\right) + (4-7)^2\left(\frac{3}{36}\right) +$$

$$(5-7)^2\left(\frac{4}{36}\right) + (6-7)^2\left(\frac{5}{36}\right) + (7-7)^2\left(\frac{6}{36}\right) +$$

$$(8-7)^2\left(\frac{5}{36}\right) + (9-7)^2\left(\frac{4}{36}\right) + (10-7)^2\left(\frac{3}{36}\right) +$$

$$(11-7)^2\left(\frac{2}{36}\right) + (12-7)^2\left(\frac{1}{36}\right) = \frac{210}{36} = 5.83\overline{3}$$

$$SD(X) = \sqrt{\text{VAR}(X)} = \sqrt{5.83\overline{3}} \doteq 2.4152$$

EXAMPLE 9. Suppose one is given the following data: 20, 28, 18, 20, 20, 36, 18, 20, 28, 18, 15, 17, 20, 14, 36, 20, 15, 28. Compute: (a) mean value (b) variance and (c) the standard deviation.

Solution. Let us assume that X is our random variable and that S has a range $\{20, 28, 18, 36, 15, 17, 14\}$, i.e., the set of all values that the data assumes. We will assign P_X for X according to the "relative frequency" with which the data occurs, i.e.,

$$P_X(20) = \frac{6}{18} \qquad P_X(15) = \frac{2}{18}$$

$$P_X(28) = \frac{3}{18} \qquad P_X(17) = \frac{1}{18}$$

$$P_{\dot{x}}(18) = \frac{3}{18} \qquad\qquad P_X(14) = \frac{1}{18}$$

$$P_X(36) = \frac{2}{18}$$

So $E(X) = \overline{X} = 20\left(\frac{6}{18}\right) + 28\left(\frac{3}{18}\right) + 18\left(\frac{3}{18}\right) + 36\left(\frac{2}{18}\right) + 15\left(\frac{2}{18}\right) +$

$$17\left(\frac{1}{18}\right) + 14\left(\frac{1}{18}\right)$$

i.e., $\overline{X} = \dfrac{20(6)+28(3)+18(3)+36(2)+15(2)+17(1)+14(1)}{18}$

$$\overline{X} = \frac{391}{18} = 21.72\overline{2}$$

Note that $\overline{X}$ is a weighted average of the data.

$$\text{Var}(X) = (20-\overline{X})^2\left(\frac{6}{18}\right) + (28-\overline{X})^2\left(\frac{3}{18}\right) + (18-\overline{X})^2\left(\frac{3}{18}\right) +$$

$$(36-\overline{X})^2\left(\frac{2}{18}\right) + (15-\overline{X})^2\left(\frac{2}{18}\right) + (17-\overline{X})^2\left(\frac{1}{18}\right) +$$

$$(14-\overline{X})^2\left(\frac{1}{18}\right)$$

$$\text{Var}(X) = 42.0895$$
$$\text{SD}(X) = 6.4876$$

The variance and standard deviation of a random variable give a measure of the "dispersion" or "spread" of the values of the random variable from the mean value of the random variable.

SELF-TEST:

1. Given the data 28, 50, 60, 50, 80, 90, 28, 50, 60, 90, 90 find $E(X)$, $\text{Var}(X)$, $\text{SD}(X)$.

ANS: $E(X) = 61.4\overline{5}$

$\text{Var}(X) = 493.157 \doteq 49$

$\text{SD}(X) = 22.207 \doteq 22$

2. Suppose two dice are rolled and the sum of the dice is recorded as a random variable, called X. Find

    (a) $P(X \geq 9)$ <u>Hint:</u> $P(X \leq 9) = P(X=2) + P(X=3)$
                                                  $+ \ldots + P(X=9)$.

    (b) $P(X=9)$

    (c) $P(X < 9)$

                                    ANS: (a) $\dfrac{5}{18}$

                                          (b) $\dfrac{1}{9}$

                                          (c) $\dfrac{13}{18}$

3. A man will win \$300 if he throws either a seven or twelve in the roll of a pair of dice. If he rolls otherwise he will lose \$50. What is the mathematical expectation for this game? Is the game fair?

                                  ANS: $E(X)$ = \$18.06
                                           No. Game is in
                                           player's favor.

EXERCISES 4.5

1. Let X be a random variable that denotes the number of tails obtained when three coins are tossed. Find E(X).

2. Let X be a random variable that denotes the number of tails obtained when three coins are tossed. Find Var(X) and SD(X).

3. Suppose a number is selected at random from the first twelve positive integers. Let X be a random variable that represents the number obtained. Find E(X).

4. Find the Var(X) and SD(X) where X is the random variable given in Problem 3.

5. Find the $P(X=2)$ for the random variable in Problem 1.

6. Find the $P(X \leq 2)$ for the random variable in Problem 1.
   Hint: $P(X \leq 2) = P(X=0) + P(X=1) + P(X=2)$.

7. Find the $P(X \leq 5)$ for the random variable in Problem 1.

8. For the random variable assigned in Example 4, find $P(X=5)$.

9. For the random variable assigned in Example 4, find $P(X \leq 5)$.

10. For the random variable assigned in Example 4, find $P(X \geq 5)$.

11. An urn contains 8 balls number 1 through 8. A ball is selected at random. Let X be the random variable which records the number on the ball. Find E(X).

12. Find Var(X) for X defined in #11.

13. Find SD(X) for X defined in #11.

247

14. For the random variable defined in #11 find $P(X \leq 4)$.

15. Suppose 40% of the voters favor a certain proposal and 60% are opposed to the proposal. Let X be a random variable such that X=1 if the person is in favor and X=0 if the person is opposed. Find $E(X)$ and $Var(X)$.

16. Suppose you buy one of 2000 tickets in a lottery where the tickets are $.25 each. The winning prize is $150. What is your expected value (net winnings) for this game?

17. Suppose the tickets in #16 cost $.40. What is the expected value?

18. A man will win $70 if he throws a total of seven in the roll of a pair of dice and will lose $12 if he does not. What is the man's mathematical expectation?

19. A man will win $100 if he throws either a seven or twelve in the roll of a pair of dice. He will lose $20 if he does not. What is the man's mathematical expectation?

20. One coin is drawn at random from a hat containing a nickel, a dime, a quarter, and a half dollar. If it cost 25¢ to make a draw, is this game fair?

21. One bill is drawn at random from a hat containing 4 one-dollar, 3 five-dollar, 2 ten-dollar and 1 twenty-dollar bills. What is a fair cost to charge for a single draw?

22. A prize of $500 is offered for winning a raffle. If the odds of winning are 1 to 250 what is a fair price to charge for a raffle ticket?

23. Given the data: 40, 68, 72, 68, 30, 40, 85, 68. Find the mean value, variance, and standard deviation.

24. Given the data: 100, 200, 300, 100, 200, 350, 100, 350, 200, 120, 300. Find the mean value, variance, and standard deviation.

25. Given the data: -3, 2, 5, -8, -4, -3, 2, -8, 2, 3, -3. Find the mean value, variance, and standard deviation.

REVIEW OF CHAPTER FOUR

1. If there are 10 members in a club, how many ways can a president, vice-president and treasurer be chosen if no person may serve in more than 1 office?

2. Evaluate   (a) P(10,6)   (b) C(10,6)

3. How many ways can a committee of four be chosen from 5 women and 6 men if at least one woman must be on the committee?

4. How many subsets does a set with 10 elements have?

5. How many arrangements can be formed from the letters of the word MATHEMATICS?

6. How many ways can 8 people be seated at a round table?

7. A straight flush consists of five cards in sequence, all cards of the same suit. However, royal flushes are excluded. An ace may be used low, before a two. How many straight flushes are possible?

8. What is the probability of getting a straight flush when 5 cards are drawn from an ordinary deck?

9. A full house consists of a pair and three of a kind in a five-card poker hand. How many full houses are possible?

10. What is the probability of obtaining a full house when 5 cards are drawn from an ordinary deck?

11. If 5 fair coins are tossed what is the probability of exactly 2 heads being obtained?

12. Suppose 5 balls are drawn simultaneously from an urn containing 5 red and 6 white balls. Find the probability that at least 3 will be red balls.

13. A blue die and a red die are tossed together. Find the probability that the sum is greater than 7, given that the blue die shows a 4.

14. Urn I contains 6 balls numbered 1 through 6. Urn II contains 7 balls numbered 1 through 7. First an urn is selected at random and then a ball is chosen. Suppose the number on the ball is odd. Find the probability that it came from Urn I.

15. Find the odds for rolling a sum greater than 8 with a pair of dice.

16. If five cards are delt from an ordinary deck find the odds that four kings are in the hand.

17. Suppose 3 machines are chosen at random from a shipment of 20 machines of which 5 are defective. What are the odds that exactly two out of the three will be defective?

18. A gambler offers 1 to 5 odds that event A will occur, and 2 to 7 odds that event B will occur. If A and B are mutually exclusive events, what odds should be given that A or B will occur?

19. Suppose 80% of the voters favor a certain proposal and 20% are opposed to the proposal. Let X be a random variable such that X = 1 if the person favors and X = 0 if the person opposes. Find E(X) and Var(X)

20. Suppose you buy one of 5000 tickets sold at $.50 each in a lottery where the prize is $250 to the winner.

What is your expected value (net winning) for this game?

21. Given the data: 42, 20, 68, 20. 90, 100, 90, 68, 68, 20. Find E(X), Var(X) and SD(X).

22. An experiment consists in drawing from a carton that contains a green, a yellow and an orange ball; then tossing a coin. What is the probability that a green ball was drawn given that the coin fell heads?

23. An experiment consists in first tossing a coin. If the coin falls heads, then a selection is made from 2 red and 3 green balls in Urn I. If the coin falls tails then a selection is made from 3 red and 2 green in Urn II. Given that the ball selected is red what is the probability that the coin fell heads?

24. In a 20 question multiple choice exam with four choices for each question, what is the probability that a student will get more than half the questions correct by randomly guessing at the answers.

25. Suppose a brother and his sister are in the same grade with 10 other students. If the children line up randomly when the bell rings, what is the probability that the brother and sister are standing next to each other?

1. Evaluate (a) C(13,3)   (b) P(8,6)

2. How many 4 element subsets has a set with 10 elements?

3. How many committees of 7 people can be chosen from a group of 5 men and 7 women if each committee must have at least 2 women as members?

4. How many ways can a 7-member committee choose a chairman, vice-chairman, and a treasurer, if no person may hold two offices?

5. How many arrangements are there of the letters of the word ARRANGEMENT?

6. Find the probability that a 5 card hand dealt from an ordinary deck has exactly three of one kind.

7. Find the probability that a 5 card hand dealt from an ordinary deck does not have jacks or better to open the game.

8. If two die are rolled and the sum recorded find the probability that the sum is 9 given that at least one of the die is an odd number.

9. If 6 coins are tossed simultaneously, find the probability that exactly 3 heads are obtained.

10. Suppose two cards are drawn from an ordinary deck of 52 cards. Find the probability that the cards are a pair given that both are red.

11. Given the data: 88, 84, 96, 84, 88, 96, 32, 96, 50,

96. 88.   Find E(X), Var(X), and SD(X).

12.   Suppose it costs a man $25 to roll a pair of dice. If
      he rolls an even sum he wins $10 or if he rolls a seven
      or eleven he wins $90. Is this a fair game? Justify
      your answer.

COMPUTER APPLICATIONS: CHAPTER FOUR

1. Write a subprogram to compute the n factorial. The
    positive integer n is to be read in.

2. Write a program to compute the number of combinations of
    n things taken k at a time. Use the factorial subprogram
    written for Problem 1. The positive integers n, k are to
    be read in. Test your program for 10 pairs of such integers.
    <u>WARNING:</u> Do not make n too large.

3. Write a program to compute the number of permutations of n
    things taken k at a time. Use the factorial subprogram
    written in Problem 1.

4. Write a program to compute the probability of drawing
    each of the following hands in a poker game:

    (a) Four of a kind.
    (b) Full house: three of a kind together with a pair.
    (c) Flush
    (d) Straight

CHAPTER FIVE

MATHEMATICS OF FINANCE

5.1 SIMPLE INTEREST

OBJECTIVES:

1. *To compute simple interest.*
2. *To define simple discount.*

Interest is considered "money charged for borrowed money." For example, money might be borrowed from a bank or loan company and a certain amount of interest charged for the use of the money over a given period of time. On the other hand, a person may invest his money in a savings account or money market certificates, and in essence the bank is borrowing the money from the investor and then pays to the investor a certain rate of interest for the use of the money. If a person does not pay off entirely his gasoline charges for a given month, then the oil company charges the person a certain interest on the remaining unpaid balance because the person is in fact borrowing the unpaid balance from the oil company.

Interest can be computed in two ways. The first is called *simple interest* in which the interest is directly proportional to the amount borrowed (or invested) and the time period over which the money is used. The money on which the interest is computed is called the *principal.*

The second way interest can be computed is called *compound interest* in which the interest is converted into principal at the end of a certain time interval and interest plus principal becomes a new principal upon which interest is then computed for the next time period. We will explore and develop formulas for computing both types of interest.

DEFINITION 5.1: *The Principal, P, is the sum of money borrowed or invested. The amount, S, is the sum to be returned or repaid at the end of a given time period.*

The principal, P, can be referred to as the *present value* of the amount, S; the amount S is called the *future value* of the principal P.

If a principal P dollars is borrowed or invested at *simple interest* at the rate R per year, the interest at the end of t years will be

$$I = PRt$$

The amount after t years will be

$$S = P + I = P + PRt = P(1+Rt)$$

Note the amount repaid is original principal plus interest.

Example 1: Mr. Barnes borrows $450 from Mr. Lewis for $3\frac{1}{2}$ years at $6\frac{3}{4}$% simple interest per year. Find the interest and the amount.

Solution: To find the interest we use I = PRt where P = $450, R = .0675 and t = 3.5 years. Hence

$$I = (450)(.0675)(3.5) = \$106.32$$

The amount S = P + I = $450 + 106.32 = $556.32.

257

SELF TEST: If a man borrows $1500 at the simple interest rate of $6\frac{1}{2}$% per year, what is the interest and amount on the loan for 6 months?

ANS: I = $48.75, S = $1548.75

Example 2: Mr. Hodges borrows $100 for one year from a loan company at 12%. He is given only $88 and is expected to repay the $100 at the end of one year. In this case we call the 12% rate the *simple discount rate*. Suppose we wish to calculate the real simple interest rate. We will use

$$I = PRt$$

where I = $12, P = $88, and t = 1 year. Hence

$$12 = (88)R(1)$$
$$\frac{12}{88} = R = 13.\overline{63}\%$$

EXERCISE 5.1

1. Find the simple interest on $300 at 6% for 8 months.
2. Find the simple interest on $2600 at 6.5% for 9 months.
3. Find the simple interest on $8300 at 5.5% for 6 months.
4. Find the simple interest on $7800 at 8% simple interest for 9 months.
5. Find the simple interest on $4230 at 7% simple interest for 6 months.
6. Find the simple interest on $2550 at 9% simple interest for 9 months.
7. How much money should be invested at 9% for one year in order to gain simple interest of $73?
8. What principal is needed to gain $100 simple interest in 4 months, if the rate is 7%?
9. What principal is needed to gain $320 simple interest in 9 months if the rate is 5.5%?
10. How much money should be invested to gain $258 simple

interest, if the rate is 12%.

11. What principal earns $968 simple interest in one year at 8%?

12. Find the principal that will amount to (principal plus simple interest) $837 in 6 months at 7%.

13. Find the principal that will amount to (principal plus simple interest) $932 in 9 months at 6.5%.

14. Mr. Denny borrows $650 for one year from a loan company. He is given only $600 and is expected to repay the $650 at the end of one year. What is the simple discount rate? What is simple interest rate?

15. John Walsh takes a student loan for $980 for one year. He is given only $910 and is expected to repay the full $980 at the end of one year. What is the simple discount rate? What is the simple interest rate?

16. Mr. Wilks borrows $2000 for one year at a simple discount rate of $11\frac{1}{4}$%. What amount of money is he actually given?

17. Mrs. Conrad borrows some money for one year at a simple discount rate of 15%. She actually receives $4230. What amount of money must she repay at the end of one year?

18. Mrs. Longacre borrows some money for one year at a simple discount rate of 12%. She actually receives $1870. What amount must she repay at the end of one year?

19. Find the actual simple interest rate for the situation in Problem 17.

20. Find the actual simple interest rate for the situation in Problem 18.

21. Develop a formula that relates the actual simple interest rate R as a function of the simple discount rate D.

5.2 COMPOUND INTEREST

OBJECTIVES:

1. To compute compound interest.
2. To define nominal and effective interest rates.
3. To compute present value.

The ordinary way that interest is computed by most savings and loan institutions is with compound interest rates. The institution computes interest at the end of set conversion periods of time. A conversion period may be a day, a month, 3 months, 6 months, 12 months or any finite amount of time. At the end of the fixed conversion period the interest is computed, and this interest is added to the current principal. Hence, principal for the next following conversion period is the principal plus interest from the subsequent conversion period. This accumulation of principal plus interest continues for the amount of time the money is invested or borrowed. The following terminology is used when referring to conversion periods.

TERMINOLOGY =	CONVERSION PERIODS/YEAR
Compounded annually	1
Compounded semi-annually	2
Compounded quarterly	4
Compounded monthly	12
Compounded daily	360 or 365

NOTE: The ordinary interest year is considered to be 360 days, while the exact interest year is considered as 365 days.

Compound interest is determined both by the annual interest rate, sometimes called the nominal interest rate, and the frequency of compounding, i.e., the number of conversion periods/year.

Suppose our nominal interest rate is 12% and the compounding is semi-annually, i.e., every six months. Then to compute interest during the first six months we use

$$I = PRt = P(.12) \frac{6 \text{ months}}{12 \text{ months}} = P(.06)$$

Thus during each six month period the interest = (.06) times the current principal. If the compounding had been quarterly then

$$I = PRt = P(.12)(\tfrac{1}{4}) = P(.03)$$

i.e., after each 3 months interest = (.03) times current principal.

Example 1: Suppose a $1000 is invested at 12% per year compounded semi-annually. What is the amount after 2 years?

Solution: Let S_i and P_i be the amount and principal respectively after i conversion periods. After the first 6 months

$$S_1 = P_1 + I_1 = 1000 + 1000(.12)(\tfrac{1}{2})$$
$$S_1 = 1000(1+.06) = \$1060$$

Now S_1 becomes P_2.

After the next 6 months

$$S_2 = P_2 + I_2 = 1060 + (1060)(.12)(\tfrac{1}{2})$$
$$S_2 = 1060(1+.06) = \$1123.60$$

Now S_2 becomes P_3.

After the next 6 months

$$S_3 = P_3 + I_3 = 1123.60 + (1123.60)(.12)(\tfrac{1}{2})$$

$$S_3 = 1123.60(1+.06) = \$1191.02$$

Now S_3 becomes P_4.

$$S_4 = P_4 + I_4$$

$$S_4 = 1191.02(L+.06) = \$1262.48$$

S_4 will be the accumulated amount after 2 years.

We now consider the general problem of finding the (accumulated) amount on an investment of P dollars for t years with a nominal rate R and the number of conversion periods in a year equal f.

$$S_1 = P + I_1 = P + PR\left(\tfrac{1}{f}\right) = P\left(1+\tfrac{R}{f}\right) = P_2$$

$$S_2 = P_2 + I_2 = P_2 + P_2 R\left(\tfrac{1}{f}\right) = P_2\left(1+\tfrac{R}{f}\right) = P\left(1+\tfrac{R}{f}\right)\left(1+\tfrac{R}{f}\right)$$

Hence $S_2 = P\left(1+\tfrac{R}{f}\right)^2 = P_3$

$$S_3 = P_3\left(1+\tfrac{R}{f}\right) = P\left(1+\tfrac{R}{f}\right)^2\left(1+\tfrac{R}{f}\right) = P\left(1+\tfrac{R}{f}\right)^3 = P_4$$

$$S_4 = P_4\left(1+\tfrac{R}{f}\right) = P\left(1+\tfrac{R}{f}\right)^3\left(1+\tfrac{R}{f}\right) = P\left(1+\tfrac{R}{f}\right)^4$$

Continuing in the above fashion we have that after n conversion periods the amount

$$S_n = P\left(1+\tfrac{R}{f}\right)^n$$

Of course the value of n, the number of conversion periods, is dependent on the number of years of the investment and the number of conversion periods/year. In fact n = ft. Our formula then becomes

$$S = P\left(1+\frac{R}{f}\right)^{ft}$$

where S is the accumulated amount after t years on a principal of P dollars invested at a nominal rate R compounded f times/ year.

Example 2: Find the amount if $500 is invested for 6 years at 8% per annum compounded

    a. annually

    b. semi-annually

    c. monthly

    d. daily

Solution:

    a. P = $500, R = 8%, t = 6, f = 1.

$$S = P\left(1+\frac{R}{f}\right)^{ft} = 500\left(1+\frac{.08}{1}\right)^{(1)(6)} \doteq \$793.44$$

    b. f now equals 2.

$$S = 500\left(1+\frac{.08}{2}\right)^{(2)(6)} = 500(1.04)^{12} \doteq \$800.52$$

    c. f = 12

$$S = 500\left(1+\frac{.08}{12}\right)^{(12)(6)} = 500(1.00\overline{6})^{72} \doteq \$806.75$$

    d. f = 360

$$S = 500\left(1+\frac{.08}{360}\right)^{(360)(6)} \doteq 500(1.000\overline{2})^{2160} \doteq \$807.99$$

SELF TEST: Find the amount at the end of 12 years if $500 is invested at 6% compounded

    a. semi-annually   ANS: $1016.40

    b. monthly   ANS: $1025.38

DEFINITION 5.2   In the formula

$$S = P\left(1+\frac{R}{6}\right)^{6t}$$

P is called the present value of the amount S. S is called the future value of P.

Example 3:   To help finance his son's college education 10 years from now, a father wishes to make an initial deposit in a trust fund paying $6\frac{1}{2}$% compounded semi-annually that will amount to \$20,000.   What sum of money should the father invest?

Solution: The question asks for the present value of P when $S = \$20,000$, ~~when~~ $R = .065$, $t = 10$, and $f = 2$.   To find P we solve the formula

$$S = P\left(1+\frac{R}{f}\right)^{ft} \text{ for } P.$$

Hence $P = S\left(1+\frac{R}{f}\right)^{-ft} = 20,000\left(1+\frac{.065}{2}\right)^{-(2)(10)}$

$$= 20,000(1.0325)^{-20} = \$10549.43$$

Thus, \$10,549.43 should be invested at this time.

DEFINITION 7.3:   The effective rate of interest is the rate of interest j which if compounded annually would yield the same interest in one year as the given nominal rate would earn in one year.

To derive a formula for this effective rate j, we first compute the amount S using the given nominal rate, i.e.,

$$S = P\left(1+\frac{R}{f}\right)^{ft} = P\left(1+\frac{R}{f}\right)^{f}$$

since t = 1.   The amount S for 1 year at the effective
interest rate j would be

$$S = P\left(1+\tfrac{j}{1}\right)^1 = P(1+j).$$

Since both amounts must be equal

$$P(1+j) = P\left(1+\tfrac{R}{f}\right)^f$$

$$1 + j = \left(1+\tfrac{R}{f}\right)^f$$

$$\boxed{j = \left(1+\tfrac{R}{f}\right)^f - 1}$$

Example 4:   What is the effective rate for a nominal rate
of 8% compounded quarterly?

Solution: R = .08, f = 4.

$$j = \left(1+\tfrac{.08}{4}\right)^4 - 1 \doteq .08243$$

Hence the effective rate is approximately 8.24%.

Example 5:   Check the result in the previous example by
computing the interest on $1000 at both the nominal and
effective rate.

Solution: At nominal rate: P = $1000, R = .08, f = 4, t = 1.
Hence.

$$S = 1000\left(1+\tfrac{.08}{4}\right)^{(4)(1)} = 1000(1.02)^4 \doteq 1082.43.$$

At the effective rate: P = $1000, j = .08243, f = 1, t = 1.

$$S = 1000(1+.08243) = 1000(1.08243) = 1082.43$$

SELF TEST: A man wishes to make an investment at this time

265

that will accumulate to $15,000 in 25 years. If the nominal
interest rate is 8% compounded quarterly, how much should
his investment be?

ANS: $2070.49

Find the effective rate for 12% compounded monthly.

ANS: 12.68%

EXERCISE 5.2

1. Find the amount and compound interest if $750 is
    invested for 8 years at 8% compounded quarterly.
2. Find the amount and compound interest if $1500 is
    invested for 10 years at 8% compounded daily.
3. Find the amount and compound interest if $1700 is
    invested for 10 years at $6\frac{1}{2}$% compounded quarterly.
4. Find the amount and compound interest if $220 is
    invested for 12 years at $6\frac{1}{2}$% compounded semi-annually.
5. Find the amount and compound interest if $4500 is
    invested at 12% for 10 years compounded monthly.
6. Find the amount and compound interest if $4500 is
    invested at 12% for 20 years compounded semi-annually.
7. Find the amount and compound interest if $2500 is
    invested at 12% compounded daily for 20 years.
8. Find the amount and compound interest if $1500 is
    invested at 12% compounded monthly for 20 years.
9. Find the compound amount and interest after 15 years
    if $8500 is invested at $7\frac{1}{2}$% compounded semi-annually.
10. Find the compound amount and interest after 25 years
    if $8500 is invested at 12% compounded monthly.
11. Find the effective rate if interest is compounded
    quarterly at a nominal rate of 6%.
12. Find the effective rate if interest is compounded
    semi-annually at a nominal rate of $6\frac{1}{2}$%.
13. Find the effective rate if interest is compounded
    annually at $6\frac{3}{4}$%.
14. Find the effective rate if interest is compounded

monthly at a nominal rate of 6%.

15. Find the effective rate if interest is compounded daily at a nominal rate of $6\frac{1}{4}$%.

16. Which is a better investment: one paying 8% per year compounded monthly or one paying 8.2% compounded semi-annually?

17. Which is a better investment: one paying $7\frac{1}{2}$% per year compounded quarterly or one paying $7\frac{1}{4}$% per year compounded daily?

18. How much should be invested now in order to have $1500 at the end of 5 years if the interest paid is a nominal rate of 8% compounded semi-annually?

19. What is the future value of $20,000 invested for 8 years at 10% per year compounded quarterly?

20. What is the future value (amount) of $20,000 invested for 12 years at 12% per year compounded quarterly?

21. What is the future value (amount) of $15000 invested for 5 years at 12% per year compounded monthly?

22. What is the future value of $15,000 invested for 5 years at 8% per year compounded monthly?

23. A deposit of $1000 will amount to $2665.80 in $12\frac{1}{2}$ years with interest compounded semi-annually. What is the nominal rate of interest?

24. A deposit of $1000 will amount to $1638.60 in 5 years. If interest is compounded semi-annually, what is the nominal rate of interest?

25. A deposit of $5,000 will amount to $10,488 in $7\frac{1}{2}$ years. Find the nominal rate of interest if interest is compounded quarterly.

26. If a deposit of $1,000 will amount to $1,343.90 in $2\frac{1}{2}$ years when interest is compounded quarterly, find the nominal rate of interest.

27. A trust fund is being set up so that a baby just born will collect $25,000 when she is 21 years old. What payment should be made if the nominal rate of interest is 9% compounded monthly?

28. A father sets up a trust fund for his new born child so

267

that when the child is 18 years old he will collect $15,000. What payment will the father make if the nominal interest rate is $9\frac{1}{2}$% compounded quarterly?

29. What nominal rate of interest compounded monthly is needed to double a principal in 5 years?

30. What nominal rate of interest compounded monthly is needed to double a principal in 7 years?

31. How long will it take $300 to amount to $500 if invested at 9% compounded monthly?

32. How long will it take $6,000 to amount to $7,500 if invested at $8\frac{3}{4}$% compounded semi-annually?

33. A major department store has a finance charge of 1% per month on the unpaid balance. What is the nominal rate compounded monthly? What is the effective rate?

34. How many years will it take for $500 to amount to $750 if the nominal rate of interest is $7\frac{1}{2}$% compounded monthly.

35. What is the nominal rate of interest compounded monthly that yields an effective rate of $9\frac{1}{4}$%?

36. What is the nominal rate of interest compounded quarterly that yields an effective rate of $6\frac{1}{2}$%?

37. How many years will it take $300 to triple at an effective rate of $8\frac{1}{4}$%?

5.3 ANNUITIES

OBJECTIVES:

1. To calculate present and future value of annuities.
2. To calculate payments for a sinking fund.

One of the most popular ways for many Americans to purchase, for example, a new color T.V. or a new automobile, is by means of installment buying. Under this type of plan a certain amount of money is paid each month together with interest on the unpaid balance. Also a popular method of saving money, say to make a down payment on a home or to take a long planned for vacation, is to have, for example, the credit union deduct from each month's pay a certain sum of money and then those monthly sums plus interest accumulate to a desired amount. Such a series of equal payments as in the above two examples is referred to as an <u>annuity</u>.

Even though the word annuity is derived from the Latin word annus meaning year, the interval of time at the end of which the payments are made can actually be any length of time. Other examples of annuities include premiums for insurance and amortizing a mortage. We will deal with what is called an ordinary annuity, i.e., one in which the payments are made at the end of the conversion periods. Slight adjustments would have to be made in our formulas if payments are to be made at the beginning of the conversion periods.

We will now define important terms that will be used in our discussion of annuities.

1. ANNUITY is defined as equal payments paid or received at regular time intervals.

2. PAYMENT PERIOD is defined as the interval of time

at the end of which payments are due.

3. TERM OF THE ANNUITY is defined as the time between the beginning of the first payment period and the end of the last.

4. RENT is defined as each payment and is denoted by K.

5. ANNUAL RENT is defined as the sum of payments in one year.

6. AMOUNT OF AN ANNUITY is defined as the total amount of money which would be accumulated at the end of the term if each rent were invested at a given rate of compound interest at the time of the payment. We will assume conversion periods of compound interest coincide with the payment period of the annuities.

7. FUTURE VALUE OF AN ANNUITY is the same as the amount of the annuity.

8. PRESENT VALUE OF AN ANNUITY is defined as the sum of the present values of all the payments at the beginning of the term or equivalently the present value which at this time would accumulate to the amount of the annuity under compound interest.

9. SINKING FUND is defined as a savings account to which equal periodic payments are made in order to have a specified amount of money accumulated at some future time.

At this point we recall an important identity from the algebra of polynomials, namely,

270

$$1 + x + x^2 + \ldots + x^{n-1} = \frac{x^n - 1}{x - 1}.$$

This identity can be informally verified by long division.

Using the above identity we now develop a formula to calculate the amount of an annuity which pays \$1 at the end of each payment period for n periods with an interest rate of R compounded f times/year period. The customary way of denoting the amount of this \$1 payment annuity is

$$s_{\overline{n}|\,i}$$

which is read "s angle n at i" where n represents the total number of conversion periods and $i = \frac{R}{f}$. We will calculate the amount by computing and summing up each accumulated amount for each \$1 payment. The amount S is computed using

$$S = P\left(1 + \frac{R}{f}\right)^{ft} = P(1+i)^n.$$

PAYMENT PERIOD	RENT (PRINCIPAL)	# OF CONVERSION PERIODS THIS RENT IS INVESTED	ACCUMULATED AMT. FOR THIS RENT
0	0	n	0
1	\$1	n-1	$(1+i)^{n-1}$
2	1	n-2	$(1+1)^{n-2}$
.			
.			
.			
.			
n-2	1	2	$(1+i)^2$
n-1	1	1	$(1+1)^1$
n	1	0	1

Therefore, the total amount is given by

$$S = 1 + (1+i)^1 + (1+i)^2 + \ldots + (1+i)^{n-2} + (1+i)^{n-1}$$

which by the polynomial identity above equals $\frac{(1+i)^n - 1}{i}$.

Hence
$$s_{\overline{n}|i} = \frac{(1+i)^n - 1}{i}$$

If the rent is changed from \$1 to \$K then the amount becomes $S = Ks_{\overline{n}|i}$.

Example 1: Donna saves \$150 every 3 months in a savings account that pays $6\frac{1}{2}$% per annum compounded quarterly. How much will Donna have saved after 10 years?

Solution: K = \$150, R = .065, f = 4, $i = \frac{.065}{4}$ and n = (10)(4) = 40. Hence

$$S = Ks_{\overline{n}|i} = 150s_{\overline{40}|\,.01625}$$

$$s_{\overline{40}|\,.01625} = \frac{(1+.01625)^{40} - 1}{.01625} \doteq 55.726693$$

Thus $S = 150s_{\overline{40}|\,.01625} \doteq 8359.00$

Example 2: Compute the amount of an annuity of \$400 invested semi-annually for 8 years at 4% per annum compounded semi-annually.

Solution: K = \$400, R = .04, f = 2, $i = \frac{.04}{2} = .02$, n = (8)(2) =

$$s_{\overline{n}|i} = s_{\overline{16}|\,.02} = \frac{(1+.02)^{16} - 1}{.02} = 18.63928$$

Hence $S = Ks_{\overline{n}|i} = 400s_{\overline{16}|\,.02} \doteq \$7455.71$

Example 3: (Sinking Funds) A man wishes to save \$23,300 for his son's future education by making a payment at the end of every six months for the next 18 years. If the interest rate is 8% per annum compounded semi-annually what rent should he pay?

<u>Solution:</u> S= $23,300, K = ?, R = .08, f = 2, $i = \frac{.08}{2} = .04$,

n = (18)(2) = 36 since

$$S = K s_{\overline{36}|.04}$$

then $\quad K = \dfrac{S}{s_{\overline{36}|.04}}$

$$s_{\overline{36}|.04} = \frac{(1+.04)^{36}-1}{.04} = 77.59831385$$

Hence $\quad K = \dfrac{23,300}{s_{\overline{36}|.04}} \doteq \$300.26$ per six months.

The present value of an annuity is found by summing up the present values of all the rents at the beginning of the term. The present value tells us what sum must be invested at the present time in order to obtain a series of equal payments beginning one conversion period later. Equivalently, the present value of the annuity is the money needed for an investment right now that will accumulate to the amount of the annuity. We use the compound interest formula

$$S = P(1+i)^n$$

to compute present value. Consider the following chart in computing the present value:

n PAYMENT PERIOD	s RENT	$\boxed{S(1+i)^{-n} = P}$ PRESENT VALUE
0	0	0
1	1	$(1+i)^{-1}$
2	1	$(1+i)^{-2}$
3	1	$(1+i)^{-3}$
$\vdots$		
n-1	1	$(1+i)^{-(n-1)}$
n	1	$(1+i)^{-n}$

Hence the present value P can be found by:

273

$$(1+i)^{-1} + (1+i)^{-2} + \ldots + (1+i)^{-(n-1)} + (1+i)^{-n} = P$$

$$(1+i)^{-n} [(1+i)^{n-1} + (1+i)^{n-2} + \ldots + 1] = P$$

$$(1+i)^{-n} \frac{(1+i)^n - 1}{(1+i) - 1} = \frac{(1+i)^n - 1}{i(1+i)^n} = P$$

Hence the present value with a rent of \$1 symbolized by

$$a_{\overline{n}|i} = \frac{(1+i)^n - 1}{i(1+i)^n} \quad \text{or} \quad a_{\overline{n}|i} = \frac{1 - (1+i)^{-n}}{i}.$$

If the rent is \$K per payment period then the present value $P = K a_{\overline{n}|i}$. We could also find the present value P by solving

$$S = K s_{\overline{n}|i} = P(1+i)^n$$

for P, i.e., finding the investment needed now to accumulate to the amount of the annuity. Hence

$$P = K \frac{s_{\overline{n}|i}}{(1+i)^n} = K a_{\overline{n}|i}$$

as before.

Example 4: Find the present value of an ordinary annuity of \$200 per month for $3\frac{1}{2}$ years at an interest rate of 6% per annum compounded monthly.

Solution: $K = \$200$, $i = \frac{.060}{12} = .005$, $n = (3\frac{1}{2})(12) = 42$. Hence

$$a_{\overline{42}|.005} = \frac{1 - (1 + .005)^{-42}}{.005} \doteq 37.7982991$$

$$P = 200 \ (37.7982991) = 7559.66$$

SELF TEST: Find the amount and present value of an ordinary annuity of \$125 for 12 years at 12% per annum compounded monthly.

ANS: S = 39,882.69

P = 9517.15

In order to further illustrate the relationship of present

274

and future value of an annuity we note that if the present
value P = 9517.15 of the annuity in the Self Test were invested
for 12 years at 12% per annum compounded monthly the accumulated amount would be S = $39,882.69, i.e.,

$$9517.15 \left(1 + \frac{.12}{12}\right)^{(12)(12)} \doteq \$39,882.69$$

EXERCISE 5.3

1. Find the amount of an annuity of $200 per year at the
   end of each year for 10 years at $6\frac{1}{2}$% per annum compounded
   annually.

2. Find the amount of an annuity of $250 per year at the
   end of each year for 15 years at $6\frac{1}{2}$% per annum compounded
   annually.

3. Find the amount of an annuity of $90 paid at the end
   of each six months for 20 years at 6% per annum compounded semi-annually.

4. Find the amount of an annuity of $50 paid at the end
   of each six months for 20 years at 6% per year compounded semi-annually.

5. If a man deposits $20 at the end of each 3 month period
   and if interest is at 7% per year compounded quarterly,
   how much money will he accumulate in 25 years?

6. If a man deposits $42 at the end of each 3 month period
   and if interest is at 7% per annum compounded quarterly,
   how much money will he accumulate in 20 years?

7. A woman deposits $25 in her savings account at the end
   of each month. The bank pays interest at $6\frac{3}{4}$% per
   annum compounded monthly. How much will she have in
   her account in 10 years?

8. A woman deposits $12 at the end of each month in a
   bank paying interest at 8% per annum compounded monthly.
   How much will she have after 5 years?

9. A boy begins a savings account on his 5th birthday by
   depositing $100 which he received from generous
   relatives. If he continues this practice through his
   25th birthday and the bank pays $7\frac{1}{2}$% interest per annum

compounded annually, how much will he have saved by
his 25th birthday?

10. Mr. Hartsell pays insurance premiums of $64.50 twice
a year. The interest is compounded semi-annually at
6% per year. What amount will the policy be worth
after 20 years?

11. Mrs. Murphy wishes to make equal deposits at the end
of each six months for the next 30 years so as to
accumulate $20,000. If the bank pays 6% per year
compounded semi-annually what must the deposit be each
six months?

12. Mr. Champayno wishes to make equal deposits at the end
of each six months so that in the next 20 years he will
accumulate $15,000. If the bank pays $6\frac{1}{2}$% per year
compounded semi-annually, what deposit must he make
at the end of each six months?

13. If the bank pays $6\frac{3}{4}$% per year compounded monthly,
what amount must be saved at the end of each month in
order to accumulate $5,000 at the end of 8 years?

14. If the bank pays 7% per year compounded quarterly,
what amount must be saved at the end of each quarter
in order to accumulate $5,000 at the end of 8 years?

15. A machine will wear out in 10 years. It is anticipated
that at that time the cost of the machine will be $9,500.
What amount must be set aside at the end of each month
in a savings account paying $6\frac{1}{2}$% per year compounded
monthly, if one wishes to have the money for a new
machine in 10 years?

16. Mr. Burch anticipates that he will need a new car
costing $7,500 in 5 years. How much should he begin
setting aside at the end of each month if the bank
pays $6\frac{3}{4}$% per year compounded monthly if he wants to
accumulate sufficient funds for the new car in 5 years?

17. When Mr. Redmond retires he wants to make a lump sum
investment paying 8% per annum compounded annually so
that he will receive an annuity payment of $6,000 per

year for the following 10 years. How much should Mr. Redmond invest as a lump sum.

18. Mrs. Frances wants to receive annuity payments of $5,000 at the end of every six months for the next 15 years. How much of a lump sum must she invest if interest is paid at 12% per annum compounded semi-annually?

19. Find the future value of $1,000 per month for 10 years at 6% per annum compounded monthly.

20. Find the future value of $860 per year for 20 years at 8% per annum compounded annually.

21. Find the present value of $1,000 per month for 10 years at 6% per annum compounded monthly.

22. Find the present value of $860 per year for 20 years at 8% per annum compounded annually.

23. In order to finance his daughter's college expenses a father wishes to make a lump sum investment paying 9% per annum compounded monthly so that his daughter will receive annuity payments of $250 per month for the next four years. How much should his lump sum investment be?

24. At the end of every 3 months $300 is invested in bonds paying $8\frac{1}{2}$% per annum compounded quarterly. What is the value of this investment after 10 years?

25. At the end of every six months $500 is invested in bonds paying 12% per annum compounded semi-annually. What is the value of this investment after 10 years.

26. Suppose that at the end of each month a certain sum of money is invested in bonds paying 12% per annum compounded monthly. If the bonds are worth $10,000 after 8 years, how much must be invested at the end of each month.

27. Suppose that at the end of every six months a certain sum of money is invested in bonds paying 12% per annum compounded semi-annually. If the bonds are worth $15,500 after 20 years, how much is invested at the

end of each six months?

28. A young couple decides that in 5 years they would like to have a down payment of $10,000 for a new house. If they have an account paying 6% per annum compounded monthly how much should they begin saving at the end of each month?

29. Suppose the payment premiums on an insurance policy are $30 every quarter and a man wishes to pay his premium for a year in one lump payment at the beginning of the year. How much should he pay if interest is at 6% per annum compounded quarterly?

30. Mr. Suarez died on December 31st and left an estate to his family of $40,000. He directed that the money be paid to them at the end of each year for the next 12 years. If the money is put in an account paying 14% per annum compounded annually, what payment will be made to them at the end of each year?

31. Determine formulas for the amount and present value of an annuity whose payments are made at the beginning of each payment period.

5.4 AMORTIZATION

OBJECTIVES:

    1. To compute an amortization schedule for loan repayment.

A loan is considered to be amortized when part of each payment is used to pay off the outstanding balance and part is used to pay off interest. Thus the loan and the interest are paid off by a series of equal payments (rent) made over equal time periods. Essentially in an amortization we wish to find the rent K which after the required payment periods at a certain percent of interest gives a present value equal to the amount of the loan. From the lenders point of reference he is investing a certain sum of money (the present value) at a given percent of interest in order to receive annuity payments of $K each payment period.

Example 1: A farmer buys several acres of land for $5,000. He agrees to amortize the amount for 4 years at 12% per annum compounded semi-annually. What will be his semi-annual payments?

<u>Solution:</u> The seller looks at this problem as an annuity problem in which by making an investment of $5,000 at 12% compounded semi-annually for 4 years he will obtain certain rent payments every six months. Hence the $5,000 is the present value of the annuity and the six month payment is the rent. Therefore,

$$P = \$5,000, \quad K = ?, \quad i = \frac{.12}{2} = .06, \quad n = (2)(4) = 8$$

$$P = Ka_{\overline{8}|.06} \quad \text{implies} \quad K = \frac{5000}{a_{\overline{8}|.06}}$$

since $\quad a_{\overline{8}|.06} = \frac{1-(1+.06)^{-8}}{.06} = 6.209793811$

Then K = $805.18 per 6 months.

We now construct an amortization schedule to illustrate Example 1.

Payment	Outstanding Balance	Interest Due	Payment Due	Equity*
1	$5000	300	805.18	505.18
2	4494.82	269.69	805.18	535.49
3	3959.33	237.56	805.18	567.62
4	3391.71	203.50	805.18	601.68
5	2790.03	167.40	805.18	637.78
6	2152.25	129.14	805.18	676.04
7	1476.21	88.57	805.18	716.61
8	759.60	45.58	805.18	759.60
Cumulative Totals	-0-	1441.44	6441.44	5000.00

*Equity indicates the part of the payment that is used toward paying off the outstanding balance.

Note that the sum of the equities equals the outstanding balance, and that the difference between the total payments and the cumulative equity equals the total interest paid.

EXERCISE 5.4

1. Mrs. Webster pays off a $20,000 mortgage, interest and principal by making equal payments at the end of each month for 18 years. Find her monthly payment if the interest is 12% per year compounded monthly.

2. Mr. Baker pays off a $28,000 mortgage interest and principal by making equal payments at the end of each month for 20 years. Find his monthly payment if the interest is 12% compounded monthly.

3. Mr. and Mrs. Byron have just purchased a new $55,000 home. They made a down payment of $10,000 and

amortized the balance for 20 years at 12% compounded monthly. What are their monthly payments?

4. What is the total payment for the house in Problem 3. How much interest is paid over the 20 year period.

5. How much (cumulative) equity will Mr. & Mrs. Byron (in Problem 3) have after 10 years?

6. How much total interest does Mr. Baker (Problem 2) pay for his mortgage?

7. How much total equity will Mr. Baker (Problem 2) have after 10 years?

8. How much total interest does Mrs. Webster (Problem 1) pay for her mortgage?

9. How much total equity does Mrs. Webster (Problem 1) have after 10 years?

10. Make a chart similar to Example 1 showing the first 2 years for Mrs. Webster (Problem 1).

11. Make a chart similar to Example 1 showing the first 2 years for Mr. Baker (Problem 2).

12. Make a chart similar to Example 1 showing the first 2 years for Mr. & Mrs. Byron (Problem 3).

13. Mr. & Mrs. Gorman bought a new home for $45,000. They made a down payment of $7,500 and amortized the balance for 25 years at 9% compounded monthly. What are their monthly payments?

14. How much do Mr. & Mrs. Gorman (Problem 13) actually pay for the home? How much interest do they pay?

15. How much equity do the Gorman's have after 5 years? 10 years? 15 years?

16. Make a chart similar to Example 1 showing the first 2 years for Mr. & Mrs. Gorman (Problem 13).

17. Mr. & Mrs. Redmond purchase an $80,000 house by making a down payment of $15,000 and amortizing the balance for 30 years at 12% compounded monthly. What will be the monthly payments?

18. What is the actual amount that the Redmonds (Problem 17) are paying for this house? How much interest are they paying?

19. How much equity will the Redmonds (Problem 17) have after

10 years? After 20 years?

20. Make a chart similar to Example 1 showing the first 2 years for the Redmonds (Problem 17).

21. The Redmonds (Problem 17) have a chance of making the same purchase and the same down payment, but amortizing the balance for 25 years at 15% compounded monthly. Find the monthly payment with this new plan.

22. Using the plan in Problem 21 how much will the house actually cost the Redmonds? How much interest will they pay?

23. Using the plan in Problem 21 how much equity will the Redmonds have after 10 years? After 20 years?

24. Make a chart similar to Example 1 to show the first 2 years for the plan in Problem 21.

REVIEW OF CHAPTER FIVE

1. Explain the difference between simple interest and compound interest.
2. What is the effective annual rate of interest?
3. Derive the compound interest formula.
4. Define an "annuity."
5. What is meant by the present value of an annuity?
6. Define "amortization."
7. Explain the idea of "equity."
8. What is a "sinking fund"?
9. How does the interest per annum differ from the interest per payment period?
10. Find the interest and the amount if $1,000 is borrowed for 5 years at 9% simple interest.
11. Find the amount accumulated in a savings deposit by making an initial lump sum payment of $750 at 8% interest per annum compounded semi-annually for 10 years.
12. Derive the formula for computing the effective annual rate of interest.
13. Find the effective annual rate of interest for 12% per annum compounded quarterly.
14. If one wished to double his investment in 10 years what interest rate compounded quarterly should one seek?
15. Donna has purchased a new car on a time payment plan of $130 per month for 36 months at 15% per annum compounded monthly. What was the cost of the car without interest?
16. Find the payment for an annuity to have a present value of $2500 after 7 years at 12% per annum compounded monthly.
17. What monthly payment is needed to pay off a loan of $20,000 amortized for 12 years at 12% per annum compounded monthly?
18. Suppose a machine will cost $10,000 in five years. How much should be set aside at the end of each month in an account paying 6% per annum compounded monthly in order to accumulate enough money in five years to buy the machine?

SAMPLE TEST: CHAPTER FIVE

1. Find the amount if $2,500 is invested for 10 years at 8% per annum compounded
    a) semi-annually
    b) monthly

2. How many years will it take for an investment to double itself if interest is paid at 8% per annum compounded semi-annually?

3. What is the effective rate of $6\frac{3}{4}$% per annum compounded quarterly?

4. Find the amount of an annuity of $150 every 6 months for 10 years at 8% per annum compounded semi-annually.

5. Find the present value of the annuity in Problem 4.

6. A man pays off a $22,000 mortgage, interest and principal, by equal payments made at the end of each month for 20 years. Find the payment if interest is paid at 12% per annum compounded monthly.

7. A machine which is projected to cost $5,000 in 5 years will wear out completely by that time. What equal payments must be made at the end of each month to pay for its replacement if interest is paid at 8% per annum compounded monthly?

8. A man pays off a $5,000 mortgage, interest and principal, by making semi-annual equal payments for 3 years. Make a schedule showing interest and equity payments.

9. Mr. Morris at age 65 is expected to live for 15 years. If he can invest at 12% per annum compounded monthly, how much should he invest to provide himself with $500 per month for the next 15 years?

10. Compute (a) $s_{\overline{48}|}.015$ (b) $a_{\overline{43}|}.015$

COMPUTER APPLICATION: CHAPTER FIVE

1. Write a program to compute the future value of an invest using the compound interest formula

    $$S = P\left(1+\frac{r}{f}\right)^n$$

2. If Manhattan Island in New York was sold for $24 in 1627, use your program to find out its value today if it had been invested at 10%, compounded annually.

3. Write a program to use the compound interest formula to find out the time required for the principal to be doubled. (Use logarithm)

4. John borrows x dollars at r% monthly interest for a fixed period of m months. The interest is calculated on the unpaid balance at the start of each month. John pays y dollars each month. Write a program to prepare a month-by-month breakdown of the loan balance and the total payment made.

5. The minimum monthly payment of a visa account is 3% of the unpaid balance. Write a program to compute the cost of making only the minimum monthly payment assuming the interest rate is 1.5% on the unpaid balance at the end of each month. Note the next month balance is computed by:

    new balance = old balance + interest-payment

6. The monthly payment P on a mortgage of amount A which will be paid off (amortized) in n years with an interest rate of r% annually is determined by the formula:

    $$P = A * \frac{r}{1200} / \left(1- \left(1+\frac{r}{1200}\right) ** (-12*n)\right)$$

    Write a program to construct an amortization table based on given A, r, n and the value P computed from above.

APPENDIX A

SOME BASIC CONCEPTS

1.1 THE SET CONCEPT

OBJECTIVES:

1. *The concept of a set is introduced.*

2. *The following terms are defined: element of a set, empty set, finite and infinite sets, subsets, proper subset and equality of sets.*

The concept of a *set* or collection of objects is very natural to our thought processes. In everyday life we all encounter or speak of sets. For example

A swarm of bees.   A school of fish.

A family of people.   A corps de ballet.

The originator of the theory of sets was the famous Danish mathematician Georg Cantor[1]. He was the first to realize

---

[1]GEORG F.L.P. CANTOR (1845-1918) was born in St. Petersburg, Russia. His father was a Danish merchant and his mother, Maria Bohm, was a talented artist. Cantor received his university education at Zurich and the University of Berlin. At Berlin his instructors were the famous mathematicians Kummer, Weierstrass, and Kronecker. He received his Ph.D. degree in 1867. Cantor had an active professional career, but it was spent at a mediocre university. He never realized

his ambition for a professorship at the University of Berlin
   At the age of 29, Cantor published his first revolutionary
paper on the theory of infinite sets. Kronecker's attack of
Cantor's work as nonsense may have contributed to Cantor's
first breakdown in 1884. Cantor was subject to recurring
nervous breakdowns throughout the rest of his life and died
in a mental institution in January 1918.

---

the importance of sets. Today sets are extensively used in

many branches of mathematics. The term *set* is undefined;

the objects in a set are referred to as *elements* of the set.

Here we shall assume that a collection of objects that is

well-defined is a set. This means that if S is a set, then

there is some rule or property that can be used to determine

whether a given object x is an element of S or is not an

element of S with no ambiguity. For instance, the collection

of all good algebra books is not well-defined. An algebra

book considered as good by one person might be considered

poor by the standards of another. Thus, such a collection

is not a set because it is not well-defined. On the other

hand, the collection of all four-legged animals is well-

defined and is a set.

   Capital letters such as A, B, C, T, Y,... will be used

to denote sets, and lower-case letters a, b, x, y,... will

be used to denote elements of a set. In symbols, we write

$x \in A$, read "x is an element of the set A." If x is not an

element of the set B, we write $x \notin B$. Thus, if N represents

the set of natural numbers, then $2 \in N$ and $125 \in N$, whereas

$-3 \notin N$.

SELF TEST: Supply the correct symbol, $\in$ or $\notin$, for each of the following statements. N is the set of all natural numbers.

(a) 5.2 ___ N (b) 0 ___ N

(c) 8 ___ N (d) $\frac{5}{3}$ ___ N

Answers: (a) $\notin$ (b) $\notin$ (c) ϵ (d) $\notin$

Clearly, a set may contain some elements such as "the set of students in this class" or no elements, such as 'the set of all women presidents of the United States prior to 1980." A set that has no elements is called an *empty set* (also called *null*, or *void set*). The empty set is designated by the symbol $\emptyset$. The empty set or a set that contains a finite number of elements is called a *finite* set. A set that is not a finite set is called an *infinite* set.

SELF TEST: Fill in each blank with *finite* or *infinite*.

(a) The set of counting numbers 1 through 10 is a(n)
_____ set.

(b) The set of all integers is a(n) _____ set.

Answers: (a) finite   (b) infinite

We use two methods to describe a set: the *tabular form* and the *set-building* form. In the tabular form, we list all the elements within braces and separate them by commas. For example, if A is the set consisting of the last three letters of the English alphabet, we write

$$A = \{x, y, z\}$$

If there are a large number of elements in a set, it is combersome to list all the elements. Thus we write

$$B = \{1,\ 2,\ 3, \ldots,\ 100\}$$

to indicate the set of natural numbers 1 through 100. Likewise we write

$$C = \{2,\ 4,\ 6,\ \ldots\}$$

to show the set of all even numbers greater than or equal to 2. Care should be exercised in using the three-dot notation. the first three elements should indicate the rule by which the remaining elements are obtained. For example, $\{2,\ 7,\ \frac{1}{4}, \ldots\}$ is not acceptable, since the rule needed to find the given elements is not evident.

The second method of designating a set, called *set-builder notation*, consists of describing an identifying property of the set. If A is a set consisting of elements x having property p, we write

$$A = \{x \mid x \text{ has property } p\}$$

which is read "the set of elements x such that x has property p." The vertical bar stands for "such that." For example, the set C of all mathematics teachers at university U can be described as

$$C = \{x \mid x \text{ is a mathematics teacher at a university } \text{th}$$

SELF TEST: Use the set-builder notation to describe the set N, the natural numbers.

N = _____.

Answer: $N = \{x \mid x \text{ is a natural number}\}$.

Suppose S is the set of all students at a university and A is the set of all students taking algebra at that university. Clearly, all the elements of A are elements of S. We say that A is a *subset* of S and we write

$$A \subseteq S$$

which is read "A is contained in S." In general, we have the following.

DEFINITION 1.1: *Set A is a subset of set B, written* $A \subseteq B$, *if every element of A is an element of B.*

The empty set $\emptyset$ is assumed to be a subset of every set. Also a nonempty set A is a subset of itself, $A \subseteq A$.

EXAMPLE 1: Let $A = \{i, o, u\}$ and $B = \{x, y, i, a, o, u\}$. Is $A \subseteq B$? Is $B \subseteq A$?

Solution: Since every element of A is also an element of B, then $A \subseteq B$. On the other hand, $x \in B$ but $x \notin A$. Therefore, B is not a subset of A, and we write $B \nsubseteq A$.

If X and Y are two sets such that $X \subseteq Y$, and Y has at least one element that is not an element of X, then X is said to be a *proper subset* of Y. We write $X \subset Y$ (notice that the horizontal bar is eliminated). Thus, in Example 1, $A \subset B$.

EXAMPLE 2: List the proper subsets of A = {a, b, c}.

Solution: {a}, {b}, {c}, {a,b}, {a,c}, {b,c}, and ∅

EXAMPLE 3: How many subsets does A = {a, b, c} have?

Solution: There are eight subsets of A. In addition to the proper subsets listed in Example 2, we would include {a, b, c}.

DEFINITION 1.2: *If the set A is a subset of the set B and the set B is a subset of the set A, then set A is equal (or identical) to set B and we write A = B.*

If set A is not equal to set B, we write $A \neq B$.

SELF TEST: Use the symbols = and ≠ to fill in the blanks.
    (a) {a, b, c} ＿＿ {c, b, a}
    (b) {2, 4, 6} ＿＿ {2, 2+2, 4+2}
    (c) {i, o, u} ＿＿ {i, u}

EXERCISES 1.1

In Exercises 1-6, state which of the given collection of objects is a set.

1. The collection of students registed in 1980 at the University of Texas.

2. The collection of tall people in this class.

3. The collection of counting numbers less than 7.

4. The collection of counting numbers greater than 34.

5. The collection of counting numbers less than 1.

6. The collection of competent mathematics teachers in this school.

In Exercises 7-12, use words to write each set notation.

7. $y \in S$ 8. $S \subseteq T$ 9. $A \not\subseteq B$

10. $x \notin Y$ 11. $\{0\}$ 12. $\emptyset$

In Exercises 13-20, state which of the given statements are true.

13. $4 \in \{1, 2, 3, 4\}$ 14. $2 \subset \{1, 2, 3, 4\}$ 15. $\{2\} \in \{1, 2$

16. $\{4, 2\} = \{2, 4\}$ 17. $\emptyset \not\subseteq \{a, b, c\}$ 18. $\emptyset = \{0\}$

19. $1 + 2 \notin \{1, 2, 3, 4\}$ 20. $\frac{4}{2} \in \{1, 2, 3\}$

In Exercises 21-35, use the listing method to represent each set that is described.

21. The set of natural numbers less than 8.

22. The set of oceans on this earth.

292

23. The set of digits less than 4.

24. The set of digits greater than 9.

25. The set of odd numbers greater than 4 and less than 12.

26. The set of presidents of the United States in 1976.

27. The set of letters in the word mathematics.

28. The set of letters in the word Mississippi.

29. The set of natural numbers greater than 10.

30. The set of all natural numbers x for which x + 6 is less than 10.

31. The set of all natural numbers x for which x + 4 is less than 8.

32. The set of all digits y for which y + 6 is less than 10.

33. The set of all digits y for which y + 8 is greater than 9.

34. The set of all natural numbers x for which x + 5 is greater than 9.

35. The set of digits y for which $\frac{30}{y}$ is a natural number.

In Exercises 36-40, use set-builder notation to describe each set.

36. {1, 3, 5, 7, 9, 11, 13}

37. {2, 4, 6, 8,...}

38. {1, 4, 9, 16, 25, ...}

39. {..., -3, -2, -1}

40. {..., -3, -2, -1, 0, 1, 2, 3, ...}

41. Which of the following is true for sets A, B, and C?

   A = {z|z is a letter of the word blow}

   B = {z|z is a letter of the word bowl}

   C = {z|z is a letter of the word bellow}

(a) $A \subset C$ (b) $A = B$ (c) $B = C$

(d) $C \subset D$ (e) $B \subseteq C$

42. List all subsets of $A = \{0, 1, 2, 3\}$.

43. If the set A has five elements, how many subsets will A have? If has six elements, how many subsets will A have?

1.2 OPERATIONS ON SETS

OBJECTIVES:

1. *Venn diagrams are used to study operations on sets.*

2. *Intersection and union of sets are defined.*

3. *Disjoint sets, universal set and complement of a set are defined.*

As a visual aid, a set S may be represented by a curve in the plane enclosing points which represent the elements of S. Thus in Figure 1, the points within the curve represent the elements r, s, t, x, y, and z of the set $S = \{r, s, t, x, y, z\}$. Note that the point representing an element p is outside the curve, since $p \notin S$.

The mathematician Leonard Eúler[1] (1707-1783) was among the first to use circles in the plane to illustrate principles of logic. Later, John Venn (1834-1923) also used diagrams to illustrate principles of logic in his book Symbolic Logic. Venn, an Englishman, was ordained into the priesthood and after twenty-four years, he resigned and devoted his efforts to the study of logic. Today diagrams such as Figure 1 are

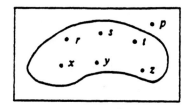

Figure 1

Set $S = \{\hbar, \delta, \ell, x, y, z\}$

are called Venn diagrams.

Operations can be performed on sets and Venn diagrams are helpful in understanding these operations. First, consider the following situation. In a group of six students (Carlos, Juanita, Debbie, Maria, Mike, and Sue), suppose the set A = {Mike, Sue, Debbie, Carlos} represents those taking a biology class and the set T = {Mike, Juanita, Maria, Carlos} represents those taking a calculus class. The set of students taking both biology and calculus is the set X = {Mike, Carlos}. The set X is derived from A and T and is called the inter-

[1] LEONARD EULER (1707-1783) was born in Switzerland. Young Leonard received his early education from his father, who had studied mathematics with Jacob Bernoulli. Euler received his university education at Basál, where he enrolled as a theology student. He received his Master's degree at the age of seventeen and wrote his first research paper at the age of nineteen. Euler was appointed to the Chair of Mathematics at the Academy of Sciences in St. Petersburg, Russia. Later he became Director of the Department of Mathematics at Berlin.

Euler was one of the leading analysts of the eighteenth century and was one of the most prolific mathematicians of all times. He wrote more than 700 papers on the various branches of mathematics. In 1735, he lost an eye and gradually became blind seventeen years before his death. He continued his work to the end. In fact, on the day he died, he was calculating the orbit of the planet Uranus.

section of A and T. We illustrate this in Figure 2.

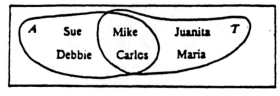

Figure 2

In general, we have the following definition.

DEFINITION 1.3: *The intersection of two sets A and B, written A ∩ B, is the set of all elements that are both in A and in B.*

For any sets A and B, we have

$$A \cap B = \{x | x \in A \text{ and } x \in B\}$$

EXAMPLE 1. Let A = {1, 2, 3} and B = {2, 4, 6, 7}. Find A ∩ B.

Solution: Since 2 ∈ A and 2 ∈ B and no other element of A has this property, then, by Definition 1.3, we have A ∩ B = {1, 2, 3} ∩ {2, 4, 6, 7} = {2}.

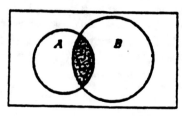
Figure 3

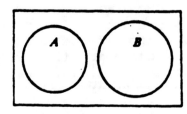
Figure 4

In the Venn diagram of Figure 3, the intersection of sets A and B is represented by the darker shaded region.

In Figure 4, we illustrate the case where A ∩ B = ∅. In such a case, we say that A and B are <u>disjoint</u> sets. For example, if A = {1, 2, 3, 4, 5} and B = {-2, -1, 0, 6}, then A ∩ B = ∅.

<u>SELF TEST:</u> Give an example of two disjoint sets.

A = _____

B = _____

A ∩ B = _____

Another operation involving sets is the union of sets.

<u>DEFINITION 1.4:</u> *The union of two sets A and B, written A ∪ B, is the set of all elements that are either in A or in B or in both. For any sets A and B, we have*

$$A \cup B = \{x \mid x \in A \text{ or } x \in B\}.$$

Figure 5 illustrates the union of the two sets A and B. Here the sets A and B are represented by the points within the respective circles. The shaded region represents the set A ∪ B.

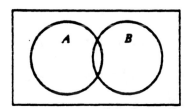

<u>Figure 5</u>   A ∪ B

297

EXAMPLE 2:   Let A = {x,y,z} and B = {i,o,x,z}. Find A ∪ B.

Solution: A ∪ B = {x,y,z} ∪ {i,o,x,z} = {i,o,x,y,z}.
Note that in A ∪ B, x and z are listed only once.

EXAMPLE 3:   If S = {a,b}, T = {b,c,d} and W = {c,d,e}, find
(S ∪ T) ∩ W.

Solution: First we find S ∪ T.

$$S ∪ T = \{a,b\} ∪ \{b,c,d\} = \{a,b,c,d\}$$

Then

$$(S ∪ T) ∩ W = \{a,b,c,d\} ∩ \{c,d,e\} = \{c,d\}.$$

SELF TEST: Let A = {0, 1, 2}, B = {0, 1}, and C = {-1. 0, 2}.
Find each of the following:

   (a)   A ∪ B = _____

   (b)   (A ∪ B) ∩ C = _____

Answers: (a)   {0, 1, 2}   (b)   {0, 2}

DEFINITION I.5:   A _universal set_, denoted by U, is a set
that contains all the elements that are being considered in
a given discussion.

    In a given discussion concerning sets, it is often neces-
sary to consider some of the elements and eliminate others.
Clearly, a universal set may vary from one discussion to
another.   Therefore, it is always important to specify the

universal set in a given discussion.

Suppose we wanted you to write a report on an animal in a zoo that does not belong to the cat family. If we asked you to write about an object that is not in the cat family, you might write about your friend, a painting by Renoir, or the Alps. To avoid that, we specify the universal set of objects for you to choose from; namely, the univeral set is the set of all animals in a given zoo. If

$$U = \{x \mid x \text{ is an animal in a given zoo}\}$$

and

$$C = \{y \mid y \text{ is in the cat family}\}$$

then you are asked to write about an element in the set

$$\overline{C} = \{z \mid z \in U \text{ and } z \notin C\}$$

The set $\overline{C}$ is called the <u>complement</u> of C.

<u>DEFINITION 1.6</u>: *The complement of a set A, denoted by $\overline{A}$, is the set of all elements in the universal set U that are not in A.*

In Figure 6, all points within the rectangle represent the universal set U and the set of all points within the circle represent the set A. The set $\overline{A}$ is the set of all points in U but not in A, that is, the shaded region

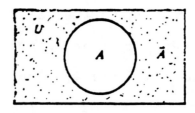

$\overline{A}$ is represented by the shaded region

Figure 6

EXAMPLE 4: U = {0, 1, 2, 3, 4, 5, 6, 7, 8, 9}

   A = {2, 4, 6}

   B = {1, 3, 5, 7, 9}

   C = {1, 2, 3, 4}

Find: (a) (A ∪ C) ∩ B   (b) (̄A ∪ B)   (c) $\overline{A}$ ∩ $\overline{B}$

Solution: (a) A ∪ C = {1.2,3,4,6}, so (A ∪ C) ∩ B = {1,3}

   (b) A ∪ B = 1.2,3,4,5,6,7,9 , so (̄A ∪ B) = {0,8}.

   (c) We have $\overline{A}$ = {0,1,3,5,7,8,9} and $\overline{B}$ = {0,2,4,6,8},
   so $\overline{A}$ ∩ $\overline{B}$ = {0,8}.

   Venn diagrams provide an easy method to illustrate operations between two or more sets. Figure 7 shows the operations (A ∩ B) ∩ $\overline{C}$. First we find A ∩ B by shading A and B. The region that is darkly shaded is the intersection of A and B. This is shown in Figure 7a. In Figure 7b, we shade A ∩ B darkly and shade $\overline{C}$ lightly. The region that is shaded in Figure 7c is (A ∩ B) ∩ $\overline{C}$. In practice, we draw only one figure and use various shades or colors to indicate the different operations.

300

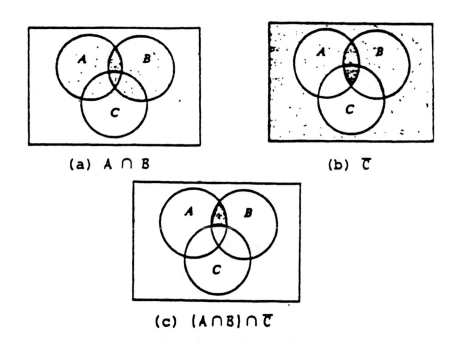

(a) A ∩ B   (b) C̄

(c) (A∩B)∩C̄

Figure 7

We now illustrate the use of sets and Venn diagrams in a practical problem.

EXAMPLE 5: A survey of 80 freshmen students revealed the following information.

55 to English	26 took English and mathematics
42 took mathematics	17 took English and psychology
23 took psychology	10 took mathematics and psychol
6 took all three subjects	

Find the number of students that took each of the following:

(a) None of the three subjects

(b) English but not mathematics or psychology

(c) Mathematics but not English or psychology

301

(d)   Psychology but not mathematics or English

Solution: In Figure 8, let U represent the set of 80 students that were surveyed.   Let E, M, and P represent the set of students taking English, mathematics, and psychology respectively.   First we place 6, the number of students who took all three subjects, in the region representing $E \cap M \cap P$.   Next, 10 students are taking mathematics and psychology, so the set $M \cap P$ must have 10 elements.   Since we already have 6 elements, we write 4 in the remainder of the region representing $M \cap P$.   Similarly, we find that 11 must be placed in the remainder of the region representing $E \cap P$ and that 20 must be placed in the remainder of the region representing $M \cap E$.   Thus, the remainder of the regions M, E, and P must contain the numbers 12, 18, and 2, respectively.

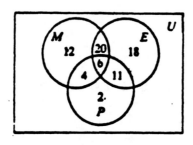

Figure 8

(a)   The total number of students taking at least one of the subjects is 73.   Therefore, there are 7 students in this survey who are taking none of the three subjects.

(b)   There are 18 students taking English but not mathematics or psychology.

302

(c) There are 12 students taking mathematics but not English or psychology.

(d) There are 2 students taking psychology but not mathematics or English.

EXERCISES 1.2

In Exercises 1-15, let U = {0,1,2,3,4,5,6,7,8,9}, A = {1,3,5}, B = {1,2,3,4,5}, and C = {4,6,8}. Determine each of the following:

1. $A \cup C$
2. $A \cap C$
3. $A \cap B$
4. $A \cap \emptyset$
5. $\overline{B}$
6. $(A \cup B) \cap C$
7. $\overline{A} \cup C$
8. $\overline{C} \cap \overline{B}$
9. $(A \cap B) \cap \overline{C}$
10. $\overline{U}$
11. $\overline{\emptyset}$
12. $A \cup \overline{U}$
13. $(\overline{A} \cap \overline{B})$
14. $(\overline{\overline{A} \cap \overline{B}})$
15. $(\overline{A} \cap \overline{B})$

16. Let A = {a,b,{a,b},c} and B = {a,b,{a,b},{a,b,c}}. Which of the following are true?

    (a) {a} $\subset$ (A $\cap$ B)
    (b) {a,b} ε (A $\cap$ B)
    (c) {a,b} $\subset$ (A $\cup$ B)
    (d) {a,c} $\subset$ A
    (e) {a,b,c} $\subset$ (A $\cup$ B)
    (f) {a,b,c} $\subset$ (A $\cap$ B)
    (g) {a,{a,b}} = A $\cap$ B
    (h) c ε (A $\cap$ B)

17. Let U = {1,a,b,c,d}. If $A \cap B$ = {a,c}, $A \cup B$ = {a,b,c,d}, $A \cap C$ = {a,b} and $A \cup D$ = {1,a,b,c}, find A, B, C and D. Is set C unique? Is set D unique?

303

In Exercises 18-21, A, B, and C are subsets of a universal set U. Use Venn diagrams to illustrate the given identities.

18. The commutative laws:

    (a) $A \cup B = B \cup A$

    (b) $A \cap B = B \cap A$

19. The distributive laws:

    (a) $A \cap (B \cup C) = (A \cap B) \cup (A \cap C)$

    (b) $A \cup (B \cap C) = (A \cup B) \cap (A \cup C)$

20. The associative laws:

    (a) $A \cup (B \cup C) = (A \cup B) \cup C$

    (b) $A \cap (B \cap C) = (A \cap B) \cap C$

21. De Morgan's[1] laws:

    (a) $(\overline{A \cup B}) = \overline{A} \cap \overline{B}$

    (b) $(\overline{A \cap B}) = \overline{A} \cup \overline{B}$

22. Draw a Venn diagram to show how each of the following sets are related. Which of the given statements (a-h) are true?

    $U = \{x \mid x \text{ is a student at the University of Texas}\}$

    $A = \{y \mid y \text{ is a student 20 years of age or older}\}$

    $B = \{z \mid z \text{ is a student under 20 years of age}\}$

    $C = \{w \mid w \text{ is a student who owns a bicycle}\}$

    (a) $A \cup B = U$ (b) $A \cup B = \emptyset$

    (c) $A \subseteq U$ (d) $A \subseteq B$

    (e) $A \cap C = \emptyset$ (f) $A \in B$

    (g) $(A \cap B) \cap C = \emptyset$ (h) $(A \cup B) \cup C = U$

---

[1]AUGUSTUS DE MORGAN (1806-1871) was a noted English mathematician and logician. He taught at University College in London and wrote several papers on the structure of algebra.

23. The <u>cardinal number</u> of a set A, denoted by $n(A)$, is the number of elements in a set A. Determine $n(A)$ for each.

(a) $A = \{a,e,i,o,u\}$

(b) $A = \{x \mid x$ is a letter of the word Mississippi$\}$

24. Give an example of a set whose cardinal number (see Exercise 23) is:

(a) 6   (b) 0

25. A survey of 350 students revealed the following information about their preference of sports.

	Golf	Cycling	Swimming	Tennis
Women	23	7	41	79
Men	32	11	29	128

Let $M = \{$men$\}$, $W = \{$women$\}$, $G = \{$golf players$\}$,
$C = \{$cyclers$\}$, $S = \{$swimmers$\}$, and $T = \{$tennis players$\}$.
Determine the cardinal number of each of the following sets (see Exercise 23).

(a) $M \cap S$ (b) $M \cup W$

(c) $M \cap T$ (d) $W \cap C$

(e) $(M \cup W) \cap G$ (f) $(M \cup W) \cap \overline{T}$

(g) $M \cap W$ (h) $(M \cup W) \cap S$

In Exercises 26 and 27, use Venn diagrams to solve.

26. In a survey of 100 freshman students the following information was obtained.

41 took mathematics

29 took psychology

27 took sociology

11 took mathematics and psychology

305

10 took sociology and psychology

6 took mathematics and sociology

4 took all three subjects

Determine the number of students that

(a) took none of the three subjects,

(b) took mathematics but not sociology or psychology,

(c) took sociology but not mathematics or psychology.

[Hint: First put in the number of students who took all

three subjects. Then determine the number of students

who took mathematics and sociology only, and so on.]

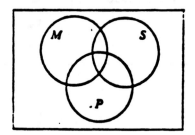

27. The Florida Theatre School has arranged a European

educational tour for 200 students.

90 are going to London

90 are going to Paris

60 are going to Rome

35 are going to London and Paris

15 are going to London and Rome

20 are going to Paris and Rome

5 are going to London, Paris and Rome

How many students are going to each of the following places?

(a) to Paris only

(b) to Rome only

(c)   to London and Paris but not Rome

(d)   to London and Rome but not Paris

(e)   to none of these cities

EXERCISES 1.1

1. $(\frac{-11}{10}, \frac{13}{10})$ 3. $(-1, 3)$ 5. $(\frac{24}{38}, \frac{-1}{38})$ 7. $(\frac{10}{16}, \frac{7}{16})$

9. no solution   11. infinite   13. $(\frac{13}{43}, \frac{59}{43})$ 15. $\simeq (-0.323, -0.657)$

17. $\sim (2.1946, 1.0878)$ 19. $(\frac{-29}{7}, \frac{-19}{7})$ 21. $(1, 2)$

23. $\sim (1.6419, 0.5209)$ 25. $(\frac{318}{121}, \frac{-37}{121})$ 27. $(0, 0)$ 33. $(\frac{16}{3}, \frac{8}{3})$

35. $(-2, 3)$ 37. $(10, 13)$ 39. 5 rab., 3 chick. 41. $(5,000, 2,000)$

EXERCISES 1.2

1. point   3. parallel 5. point   7. same line   9. point

11. point   15. $(4,000, 1,000)$ 17. $(80, 40)$

19. $(300, 100)$ 21. $(70¢, 90¢)$ 23. $(75¢, 35¢)$ 25. $(100, 200)$

27. $(3.00, .75)$ 29. no solution

EXERCISES 1.3

1. $(13/5, -4/5, -3)$ 3. $(1, 0, -1)$ 5. $(\frac{k+1}{7}, \frac{10k-4}{7}, k)$

7. no solution   9. $(\frac{910}{313}, \frac{81}{313}, \frac{194}{313}, \frac{247}{313})$ 11. $(0, 0, 0)$

13. $(k, 0, -k)$ 15. $(2+w-z, \frac{-5-w-2z}{3}, w, z)$ 17. $(1+k, -2k, k)$

19. $(1-r-s, r, s)$ 21. $(-2w, z+w-1, w, z)$ 23. $(1, 1, -1, 0, 2)$

25. $(\sim 0.540, \sim 7.004, \sim 2.195)$ 29. $(45, 45/2, 45/2)$ 31. $(3.50, 1.25, 0.75)$

33. $(3, 7, 5)$ 35. $(20, 15, 5)$ 39. Joe is 34, Mike is 31, Jane is 27,

and Mother is 59.

CHAPTER TEST 1

1a. $(\frac{10}{17}, \frac{-1}{17})$ 1b. $(\frac{11}{25}, \frac{41}{23})$ 2a. point   2b. parallel lines

3(a) $(2, 3, -5)$ 3(b) $(\frac{-7-z-3w}{7}, \frac{14 + 11z + 5w}{7}, z, w$ 3(c) no solution

3(d) $(0, -3/4r, -1/4r, r)$ 4. line   5. (5 rabbits, 7 chickens)

6. $(1200, 800)$ 7. $(1.57, 1.82, 3.55)$ 8. $(8, 7, 31, 24$ costs $1138,

$(7, 8, 31, 24$ costs $1130)$

1(a) 3×4   1(b) 2×3   1(c) 1 and 2   1(d) 2 and 9

2(a) 3+4=7   2(b) 1=y-1   3(a) m×n   3(b) $c_{4,3}$ 3(c) $c_{r,s}$

4. $\begin{bmatrix} 3 & 6 \\ -3 & 3 \end{bmatrix}$ 5. $\begin{bmatrix} 5 & 2 & 2 \\ 2 & 3 & 0 \\ 3 & 1 & 2 \\ 4 & 5 & 4 \end{bmatrix}$ 6. 5   7. 1, ‾5, 1

8. b_{11}, b_{22}, b_{33}, ..., b_{nn} 9. $a_{32} = a_{23}$ 10. any diagonal matrix

11. any diagonal matrix with some unequal elements.   12. 0   13. $\begin{bmatrix} 1 \\ 2 \\ 3 \end{bmatrix}$

14. $I_2 = \begin{bmatrix} 1 & 0 \\ 0 & 1 \end{bmatrix}$ $I_3 = \begin{bmatrix} 1 & 0 & 0 \\ 0 & 1 & 0 \\ 0 & 0 & 1 \end{bmatrix}$ $I_4 = \begin{bmatrix} 1 & 0 & 0 & 0 \\ 0 & 1 & 0 & 0 \\ 0 & 0 & 1 & 0 \\ 0 & 0 & 0 & 1 \end{bmatrix}$

15. 5 and 0

EXERCISES 2.2

1. x = 7
   y = 2

3. $x = \sqrt{2}$ 5. (a) $\begin{bmatrix} 9 & -3 \\ 6 & 18 \\ 21 & 27 \end{bmatrix}$ (c) $\begin{bmatrix} -3 & 1 \\ -2 & -6 \\ -7 & -9 \end{bmatrix}$

6. x = -1   9. $A \cdot I = A$

11. (a) $\begin{bmatrix} 2 & 4 & 5 \\ -3 & 10 & 1 \\ 5 & 10 & -2 \end{bmatrix}$, (b) = (a) , (c) $\begin{bmatrix} -1 & 17 & 2 \\ -7 & 26 & -10 \\ -2 & 15 & 12 \end{bmatrix}$ (d) $\begin{bmatrix} 13 & 24 & 1 \\ -7 & 17 & 2 \\ -14 & 33 & 7 \end{bmatrix}$ (e) =(c)

13. $\begin{bmatrix} 5 & 13 \\ 0 & 0 \\ 2 & 7 \end{bmatrix}$ 15. (a) The kth row of A•B has all zero elements.

EXERCISES 2.3

1. (a) $\begin{bmatrix} 1 & 3 & -5 \\ 0 & 5 & 6 \\ 2 & 1 & 4 \end{bmatrix}$ (b) $\begin{bmatrix} 4 & 12 & -20 \\ 2 & 1 & 4 \\ 0 & 5 & 6 \end{bmatrix}$ (c) $\begin{bmatrix} 1 & 3 & -5 \\ 2 & -14 & -14 \\ 0 & 5 & 6 \end{bmatrix}$

3. (a) $\begin{bmatrix} +1 & -1 & 3 \\ -6 & 0 & -12 \\ 5 & -1 & 1 \end{bmatrix}$ (c) C = D   5. a, b, e, f are row reduced

7. (a) $\begin{bmatrix} 1 & -1 \\ 2 & -3 \end{bmatrix}$ (b) no   (c) $\begin{bmatrix} 1 & -5/9 \\ 0 & 1/9 \end{bmatrix}$ (d) $\begin{bmatrix} 3 & 5 \\ -1 & -2 \end{bmatrix}$

9. Invertible if x ≠ 2 or x ≠ 3

11. (a) $\begin{bmatrix} 0 & 1 & 0 \\ 1 & 0 & 0 \\ 0 & 0 & 1 \end{bmatrix}$ (b) $\begin{bmatrix} 1 & 0 & 0 & 0 \\ 0 & \frac{1}{2} & 0 & 0 \\ 0 & 0 & 1 & 0 \\ 0 & 0 & 0 & 1 \end{bmatrix}$ (c) $\begin{bmatrix} 1 & 0 & 0 \\ 3 & 1 & 0 \\ 0 & 0 & 1 \end{bmatrix}$

13. $A = \begin{bmatrix} 1 & 2 \\ 2 & 4 \end{bmatrix}$

$B = \begin{bmatrix} 2 & 6 \\ -1 & -3 \end{bmatrix}$

EXERCISES 2.4

1. $\begin{bmatrix} 2 & 3 & -5 \\ 1 & -4 & 6 \\ 2 & 2 & -6 \end{bmatrix} \begin{bmatrix} x_1 \\ x_2 \\ x_3 \end{bmatrix} = \begin{bmatrix} -2 \\ -1 \\ 5 \end{bmatrix}$ 3. $\begin{bmatrix} 6 & -3 & 4 & -2 \\ 2 & 2 & -1 & 5 \\ -12 & 0 & 1 & -5 \end{bmatrix} \begin{bmatrix} x_1 \\ x_2 \\ x_3 \\ x_4 \end{bmatrix} = \begin{bmatrix} -1 \\ 2 \\ 3 \end{bmatrix}$

5. $\begin{cases} -x_1 + 2x_2 \qquad\qquad + 3x_4 = 0.1 \\ 15x_1 + 7x_2 + 4x_3 + 9x_4 = -0.5 \\ -5x_1 + 6x_2 + 0.8x_3 + 7x_4 = -1.9 \end{cases}$ 7. $x_1 = -1/5$, $x_2 = 4/5$, $x_3 = 2/5$

9. no solution

11. $x = 0$, $y = 1$, $z = 0$

13. infinitely many solutions

15. infinitely many solutions

$\begin{bmatrix} -1 & 2 & 0 & 3 \\ 15 & 7 & 4 & 9 \\ -5 & 6 & 0.8 & 7 \end{bmatrix} \begin{bmatrix} x_1 \\ x_2 \\ x_3 \\ x_4 \end{bmatrix} = \begin{bmatrix} 0.1 \\ -0.5 \\ -1.9 \end{bmatrix}$

CHAPTER 2 TEST

1. $\begin{bmatrix} 1 & 0 & 0 \\ 0 & 2 & 0 \\ 0 & 0 & 3 \end{bmatrix}$ 2. (a) $\begin{bmatrix} 8 & 6 \\ 4 & 11 \end{bmatrix}$ (b) $\begin{bmatrix} -1 & 9 & 4 \\ 1 & 11 & 11 \end{bmatrix}$ (c) not possible

(d) $\begin{bmatrix} 2 & 3 & -3 \\ 16 & 4 & 6 \end{bmatrix}$ 3. $\begin{bmatrix} 1 & 0 & 0 & 1 \\ 0 & 1 & 0 & 3 \\ 0 & 0 & 1 & 5 \end{bmatrix}$ 4. $x = 2$ or -1

5. $\begin{bmatrix} 1/4 & 1/2 & 1/4 \\ -1/2 & 1 & 1/2 \\ -1/4 & 1/2 & 3/4 \end{bmatrix}$ 6. $\begin{bmatrix} 1 & 0 \\ 0 & 1 \end{bmatrix} \begin{bmatrix} 1 & k \\ 0 & 0 \end{bmatrix} \begin{bmatrix} 0 & 1 \\ 0 & 0 \end{bmatrix}$

7. Row 2 is interchanged with Row 3, then $(-1)R_3$ is added to Row 2.

8. $x = 57 - 22w$, $y = 8w - 21$, $z = 4 - 2w$

9. If a matrix is invertible, it can be reduced to I.

10. (a) $A = \begin{bmatrix} 1 & 3 \\ 3 & 9 \end{bmatrix}$ $B = \begin{bmatrix} -6 & -9 \\ 2 & 3 \end{bmatrix}$ (b) $\begin{bmatrix} 1 & 0 & 0 \\ 0 & 1 & 0 \end{bmatrix}$ and $\begin{bmatrix} 1 & 0 \\ 0 & 1 \\ 0 & 0 \end{bmatrix}$

1. Max is 8 at (2, 0)   3. Max is 10 at $(0, \frac{10}{3})$

5. Max is 35 at (5, 1)   7. Min is 32 at (4, 0)

9. Min is 20 at (2, 4), (10, 0)   11. Max is 15 at (0, 3), min is 2 at (2, 0)

13. Max is 375 at (5, 20), (15, 10), min is 0 at (0, 0)

15. $9000 in A and $3000 in B.

17. 300 of A and 125 of B   19. 70 to I, 55 to II, C = $1765

EXERCISES 3.3

1. Max is 4 at (3, 1)   3. Max is 60 at (20, 0, 0)

5. Max is 125 at (0, 0, 0, 25)   7. Min is 5/2 at (0, 5/2)

9. Min is $\frac{21}{5}$ at $(\frac{8}{5}, 0, \frac{13}{5})$

EXERCISES 3.4

From F_1 to w_1 = 500, from F_2 to w_1 = 300, to w_2 = 100, to w_3 = 200

CHAPTER 3 TEST

1. (a) Max = $\frac{61}{7}$ at $(\frac{8}{7}, \frac{9}{7})$ (b) Min = 30 on x + 2y = 10 between (0,5) and (10,0)

(c) Max = 180 at (20, 20)   2. Max = 30 at (0, 10, 0)

(b) Min = $\frac{20}{3}$ at $(\frac{20}{3}, 0, 0)$

EXERCISES 4.1

1. 17576000   3. 120   5. 6   7. 12   9. 6840   11. 504

14. 36   17. 8   19. 1024   21. 36   23. 725760   25. 320

EXERCISES 4.2

1. (a) 1680   (b) 70   3. (a) 504   (b) 84

5. 504   7. 840   9. 720   11. 45360   13. 720   15. 220

17. 11760   19. none   21. 16   23. 2,598960   25. 4512

27. 280   29. 3045

EXERCISES 4.3

1. $\frac{1}{6}$ 3. $\frac{15}{36}$ 5. $\frac{1}{4}$ 7. 1287/2,598,960   9. 4512/2,598,960

11. 1302540/2,598,960   13. 4/8   15. 2/5   17. 7/8   19. 1/3

21. 40/126   23. 121/126   25. 1/8   27. 1/3   29. 624/2598336

EXERCISES 4.3 cont'd

31. 82/91

EXERCISES 4.4

1. 4/6   3. 1/9   5. 1/2   7. 1/11   9. 1/22   11. 6/130

13. 1/2   15. 1/2   17. 2/5   19. 712842/6497400   21. 3/35

23. .145   25. (a) .9903   (b) .9260   (c) .9510   27. 8/13

29. 1/7

EXERCISES 4.5

1. 1.5   3. 6.5   5. 3/8   7. 1   9. 5/18   11. 4.5

13. 2.29   15. 0.40 and 0.24   17. -$0.325   19. $3.33 1/3

21. $5.90   23. 58.875, 331.360, 18.203   25. -1.36, 17.87, 4.23

CHAPTER 4 TEST

1. (a) 286   (b) 28   2. 210   3. 792   4. 35   5. 4,989,600

6. 54912/2598960   7. 0.702   8. 4/27   9. 5/16   10. 26/650

11. 81,636 , 458.667, 21.417

EXERCISES 5.1

1. 12   3. $228.25   5. $148.05   7. $811.11   9. $7757.58

11. $12,100   13. $888.68   15. 7.14%, 7.69%   17. $4976.47

19. 17.65%

EXERCISES 5.2

1. 1413.41, 663.41   3. 3239.45, 1539.45   5. 14851.74, 10351.74

7. 27546.87, 25046.87   9. 25648.51, 17148.51   11. 6.14%

13. 6.75%   15. 6.45%   17. 7.71%, 7.52%   19. 44075.14

21. 27250.45   23. 8%   25. 10%   27. 3803.52   29. 13.94%

31. 69 mos.   33. 12.68%   35. 8.88%   37. 13.86 yrs.

EXERCISES 5.3

1. 2698.88   3. 6786.11   5. 5335.04   7. 4268.10

9. 4755.25   11. 122.66   13. 39.43   15. 56.42   17. 40260.49

19. 163879.34   21. 90073.45   23. 10046.20   25. 18392.80

EXERCISES 5.3 cont'd

27. 100.16   29. 115.63

EXERCISES 5.4

1. 226.39   3. 495.49   5.20464.46   7. 21489.91   9. 6070.74

13. 314.70   15. 17522.94, 21473.05, 27657.67   17. 668.60

19. 34278.72, 48400.15   21. 986.33   23. 33753.50, 56159.28

CHAPTER 5 TEST

1. 5477.81, 5549.10   2. 8.84 yrs.   3. 6.92%   4. 4466.71

5. 2038.55   6. 315.64   7. 68.25   9. 41660.83

10. 69.56522, 34.042554

CHAPTER 8

TRIGONOMETRY

8.1 Ratios in Right Triangles.

OBJECTIVE:

To define the trigonometric functions as ratios of
corresponding sides of similar right triangles.

Trigonometry is primarily based on the geometric theorem which states

that the lengths of corresponding sides of similar triangles are proportional.

8.1.1 Theorem. Two triangles are similar, denoted "$\sim$", if their correspondir

angles are equal.

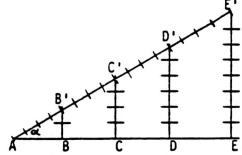

Marks on triangles are units of measurement.

$$\triangle ABB' \sim \triangle ACC' \sim \triangle ADD' \sim \triangle AEE'$$

With α as the given angle, note the ratio of the length of the opposite

side to the length of the hypotenuse in each triangle is:

$$\frac{BB'}{AB'} = \frac{2}{4} = 0.5 \qquad \frac{DD'}{AD'} = \frac{6}{12} = 0.5$$

$$\frac{CC'}{AC'} = \frac{4}{8} = 0.5 \qquad \frac{EE'}{AE'} = \frac{8}{16} = 0.5$$

8.1.2 Definition. Let α represent one of the acute angles of a right triangle.

$$\text{sine}\,\alpha = \sin\alpha = \frac{\text{side opposite}\,\alpha}{\text{hypotenuse}}$$

8.1.3 Example. Given a triangle in which $\sin\alpha = 0.5$ and the hypotenuse c = 50 ft. Find a .

Solution: $\dfrac{a}{c} = \sin\alpha$

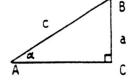

$\dfrac{a}{50} = 0.5$

$a = 50(0.5) = 25$ ft.

SELF QUIZ. Given triangle ACB as shown above in which $\sin\alpha = 0.70$ and c = 100 meters. Find a.

ANS. a = 70 meters

The decimal values found in tables and by use of calculators usually are decimal approximations of the ratios of the sides of right triangles changed to a decimal.

8.1.4 Example. Given: $\sin\alpha = 0.400$. Draw a picture of a triangle that illustrates this ratio.

Solution: $\sin\alpha = 0.400$ is equivalent to $\sin\alpha = \dfrac{0.4}{1}$. Since

$\sin\alpha = \dfrac{\text{side opposite}\,\alpha}{\text{hypotenuse}}$ a graphical illustration

could be:

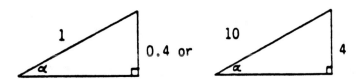

SELF QUIZ. Sketch a right triangle containing $\angle\alpha$.

1. Given: $\sin\alpha = 0.800$

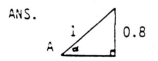

ANS.

2. Given: $\sin \alpha = 0.300$

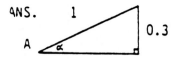

ANS.

Next consider the ratio: $\dfrac{\text{side adjacent to } \alpha}{\text{hypotenuse}}$

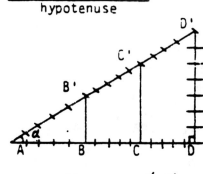

adjacent to ∡ α

By measurement:

$\dfrac{AB}{AB'} = \dfrac{4.5}{5} = 0.9$

$\dfrac{AC}{AC'} = \dfrac{9}{10} = 0.9$

$\dfrac{AD}{AD'} = \dfrac{13.5}{15} = 0.9$

Note that the ratios of the $\dfrac{\text{side adjacent}}{\text{hypotenuse}}$ = same value for the three

triangles. This ratio holds true because the triangles are similar.

<u>8.1.5 Definition</u>. Let α represent one of the acute angles of a right

triangle.

$$\text{cosine } \alpha = \cos \alpha = \dfrac{\text{side adjacent } \alpha}{\text{hypotenuse}}$$

<u>8.1.6 Example</u>. Given: $\cos 36.9° = 0.80$ and the sketch below. Find the

length of c.

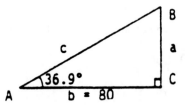

316

Solution: $\dfrac{b}{c}$ = cos 36.9°

$\dfrac{80}{c}$ = $\dfrac{0.8}{1}$

$\dfrac{80}{.80}$ = c

100 = c

SELF QUIZ. Given: cos 60° = 0.5 and b = 80 ft. as shown in the diagram

below. Find c.

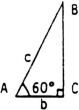

ANS. c = 160 ft.

Next consider the ratio: $\dfrac{\text{side opposite } \alpha}{\text{side adjacent to } \alpha}$

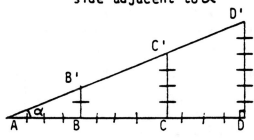

By measuring: $\dfrac{BB'}{AB}$ = $\dfrac{CC'}{AC}$ = $\dfrac{DD'}{AD}$ = 0.5

8.1.7 Definition. Let α represent one of the acute angles of a right

triangle.

tangent α = tan α = $\dfrac{\text{opposite side}}{\text{adjacent side}}$

8.1.8 Example. Given: tan α = 2.00. Sketch a related triangle.

Solution: tan α = $\dfrac{2}{1}$

.The reciprocal functions will not be emphasized in this introduction

to trigonometry but it might be noted that:

317

$$\text{cotangent}\,\Theta = \frac{1}{\text{tangent}\,\Theta} \quad,\quad \tan\Theta \neq 0$$

$$\text{secant}\,\Theta = \frac{1}{\text{cosine}\,\Theta} \quad,\quad \cos\Theta \neq 0$$

$$\text{cosecant}\,\Theta = \frac{1}{\text{sine}\,\Theta} \quad,\quad \sin\Theta \neq 0$$

The sine, cosine and tangent will be referred to as the three principal trigonometric functions.

<u>8.1.9 Example</u>. Determine the three principal trigonometric functions of Θ (theta) in the figure below. (Give an approximation to the final answers to four significant digits.)

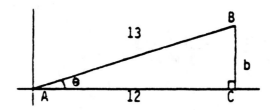

<u>Solution:</u> First apply the Pythagorean Theorem to find b.

From $c^2 = a^2 + b^2$

$b = \sqrt{c^2 - a^2}$

$b = \sqrt{13^2 - 12^2}$

$b = \sqrt{169 - 144}$

$b = \sqrt{25}$

$b = 5$

Applying the definitions of sine, cosine and tangent:

$\sin\Theta = \dfrac{\text{opposite side}}{\text{hypotenuse}} = \dfrac{5}{13} = 0.3846$

$\cos\Theta = \dfrac{\text{adjacent side}}{\text{hypotenuse}} = \dfrac{12}{13} = 0.9231$

$\tan\Theta = \dfrac{\text{opposite side}}{\text{adjacent side}} = \dfrac{5}{12} = 0.4167$

318

8.1.10 **Example.** Given the standard right triangle:

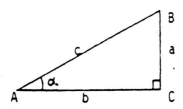

Determine the six trigonometric functions of $\angle \alpha$.

Solution: Reciprocal Functions

$$\sin\alpha = \frac{a}{c} \qquad\qquad \csc\alpha = \frac{c}{a}$$

$$\cos\alpha = \frac{b}{c} \qquad\qquad \sec\alpha = \frac{c}{b}$$

$$\tan\alpha = \frac{a}{b} \qquad\qquad \cot\alpha = \frac{b}{a}$$

SELF QUIZ. Let $\angle \beta$ be the reference angle of the standard right triangle and write the six trigonometric functions of $\angle \beta$.

$$\text{ANS. } \sin\beta = \frac{b}{c} \qquad\qquad \csc\beta = \frac{c}{b}$$

$$\cos\beta = \frac{a}{c} \qquad\qquad \sec\beta = \frac{c}{a}$$

$$\tan\beta = \frac{b}{a} \qquad\qquad \cot\beta = \frac{a}{b}$$

An angle that has its vertex at the origin of a coordinate system and its initial side along the positive x axis is said to be in standard position.

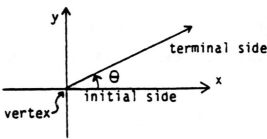

8.1.11 **Example.** Given the point (8,6). Determine the six trigonometric functions of θ (express final answers to four significant digits).

319

Solution: Locate the point (8,6). Form
a right triangle as shown in
the illustration.

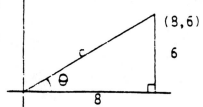

$c^2 = 8^2 + 6^2$

$c = 64 + 36$

$c = 100 = 10$

Then, $\sin \Theta = \dfrac{6}{10} = 0.6000$

$\cos \Theta = \dfrac{8}{10} = 0.8000$

$\tan \Theta = \dfrac{6}{8} = 0.7500$

$\cot \Theta = \dfrac{8}{6} = 1.333$

$\sec \Theta = \dfrac{10}{8} = 1.250$

$\csc \Theta = \dfrac{10}{6} = 1.667$

SELF QUIZ. Given the point (7,24). Determine the six trigonometric functions of Θ. (Express final answers to 4 significant digits.)

ANS. $\sin \Theta = 0.9600$ $\cot \Theta = 0.2917$

$\cos \Theta = 0.2800$ $\sec \Theta = 3.571$

$\tan \Theta = 3.429$ $\csc \Theta = 1.042$

8.1.12 Definition. If the sum of two acute angles is 90°, then they are said to be complementary; either is called the complement of the other.

8.1.13 Example. The following pairs of acute angles are complements of
each other: 30° and 60°

25° and 65°

15° and 75°

The prefix "co-" in the words cosine, cosecant and cotangent is the

320

abbreviated form of the word complement. Note the following relationships.

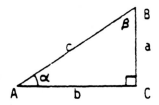

Since $\sin\alpha = \dfrac{a}{c}$ and $\cos\beta = \dfrac{a}{c}$, $\sin\alpha = \cos\beta$. In a similar manner it can

be shown that $\sec\alpha = \csc\beta$ and $\tan\alpha = \cot\beta$.

8.1.14 Example.

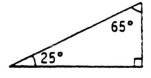

$$\sin 25° = \cos 65°$$

$$\sec 25° = \csc 65°$$

$$\tan 25° = \cot 65°$$

SELF QUIZ. Solve for Θ.

1. $\sin 27° = \cos \Theta$

ANS. 63°

2. $\tan 76° = \cot \Theta$

ANS. 14°

3. $\sec 52° = \csc \Theta$

ANS. 38°

8.1 Exercise Set.

1. Sketch a right triangle containing $\angle \Theta$.

(a) $\sin \Theta = 0.800$

(b) $\cos \Theta = 0.500$

(c) $\tan \Theta = 1.000$

(d) $\cos\theta = 0.625$

2. Given the coordinates of a point. Determine the three principal
    trigonometric functions of the angles whose terminal side passes
    through the point. (Express final answers to four significant digits.)

    (a) (6,8)   (c) (4.3,4.2)

    (b) (5,12)   (d) (11,12)

Evaluate the trigonometric functions of α and β from the given right
    triangles. (Give answers to four significant digits.)

3.

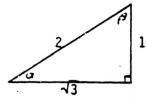

6.

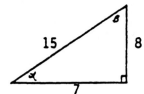

4.

7.

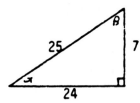

5.

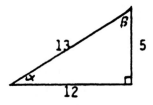

8.

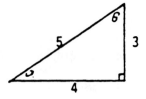

In problem 9 - 14, determine the six
    trigonometric functions of α
    and β .

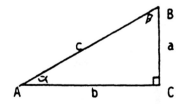

322

9. a = 6, b = 8   12. a = 5, b = 2

10. a = 24, c = 25   13. a = 24, b = 7

11. b = 8, c = 17   14. a = 12, b = 5

Draw a right triangle with the acute angle Θ and find Θ's other trigonometric

functions (express answers to 4 significant digits).

15. $\sin \Theta = \dfrac{4}{5}$ 20. $\sec \Theta = 5$

16. $\sec \Theta = \dfrac{13}{12}$ 21. $\cot \Theta = \dfrac{7}{24}$

17. $\cot \Theta = \dfrac{5}{12}$ 22. $\csc \Theta = \dfrac{2}{3}$

18. $\cos \Theta = \dfrac{1}{2}$ 23. $\sin \Theta = 0.2500$

19. $\tan \Theta = \dfrac{4}{3}$ 24. $\tan \Theta = 2.500$

Given the coordinates of a point. Determine the trigonometric functions

of the angles in standard position whose terminal sides pass through

the given points. Give results to four significant digits.

25. (8,6)   26. (7,24)   27. (8,15)

28. (3.5,7.4)   29. (11,12)   30. (13,5)

323

Find θ.

31. $\sin 50° = \cos \theta$

32. $\sec 35° = \csc \theta$

33. $\cos 40° = \sin \theta$

34. $\cot 25° = \tan \theta$

35. $\tan 57° = \cot \theta$

36. $\csc 75° = \sec \theta$

8.2 Solution of Right Triangles.

The standard way of labeling a right triangle is given below.

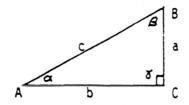

Each side is named by the lower case letter of the angle to which it is opposite. Solving a right triangle means to find certain unknown parts in the triangle. In solving for an unknown element, select the appropriate trigonometric ratio.

Use trigonometric formulas to find the unknown sides. As a check of the solution obtained apply the Pythagorean Theorem as follows:

Does $a^2 + b^2 = c^2$?

Keep in mind that approximate values will not result in (exact) equality.

If $\measuredangle \alpha$ and $\measuredangle \beta$ are both unknowns we have to resort to an inverse function.

8.2.1 Definition. If $\sin \alpha = \dfrac{a}{c}$, then $\alpha = \text{Arcsin } \dfrac{a}{c}$, read "alpha equals the acute angle whose sine is $\dfrac{a}{c}$."

8.2.2 Definition. If $\cos \alpha = \dfrac{b}{c}$, then $\alpha = \text{Arccos } \dfrac{b}{c}$.

8.2.3 Definition. If $\tan \alpha = \dfrac{a}{b}$, then $\alpha = \text{Arctan } \dfrac{a}{b}$.

The notations Arcsin, Arccos and Arctan are sometimes written as Sin^{-1}, Cos^{-1}, Tan^{-1}. These are merely shorthand notations; the meaning remains the same.

8.2.4 __Example__. Solve the following triangle by using trigonometric functions. Check your answers by using the Pythagorean Theorem.

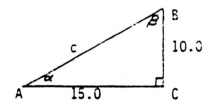

Answers should be to three significant digits.

__Solution__: α = Arctan $\dfrac{10}{15}$

α = Arctan 0.6667

α = 33.7°

β = 90° - 33.7°

β = 56.3°

The sine or cosine function may be used to find c.

By definition of sine,

sin 33.7° = $\dfrac{10}{c}$

c $\doteq$ $\dfrac{10}{sin\ 33.7°}$

c $\doteq$ 18.0

Check: $15^2 + 10^2 \doteq 18^2$?

$\qquad$ 325 $\doteq$ 324

The result 325 $\doteq$ 324 is sufficient to accept the solution as approximately correct.

<u>SELF QUIZ</u>. Solve the given right triangle.

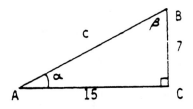

ANS. $\alpha = 25.0°$, $\beta = 65.0°$, c = 16.6

The expressions "angle of elevation" and "angle of depression" sometimes occur in word problems.

<u>8.2.5 Definition</u>. The angle of elevation is the angle formed by the "line of sight" and the horizontal ray as illustrated.

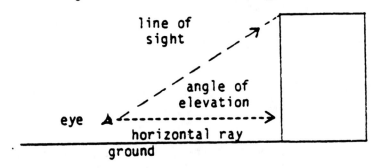

<u>8.2.6 Definition</u>. The angle of depression is the angle formed by the "line of sight" and the horizontal ray.

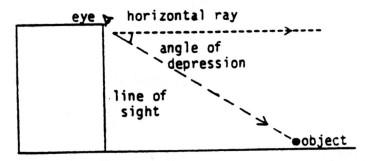

It is suggested that for the word problems a sketch be made in accordance with the given information. The picture will be an aid in analyzing the problem.

8.2.7 Example. The length of string on which a kite is flying is 500 ft. The string makes an angle of 50° with the ground. Assuming that the string is taut, how high is the kite?

Solution: Usually, the height of the person is neglected ir this type of problem.

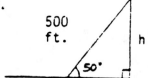

Applying the definition of the sine function:

$\sin 50° = \dfrac{\text{opposite side}}{\text{hypotenuse}}$

$\dfrac{h}{500} = \sin 50°$

$h = 500 \sin 50°$

$h = 383 \text{ ft.}$

SELF QUIZ. The pilot of a plane notes that the angle of depression to the airport is 15°. If the altimeter indicates that the altitude of the plane is 7600 ft., what is the ground distance in miles from the airport?

ANS. 5.37 miles.

8.2 Exercise Set.

Solve the right triangle. Values are associated with the standard right triangle. Answers should be expressed to three significant digits.

1. $\alpha = 40.2°$, c = 15.0

2. $\alpha = 35.5°$, b = 12.5

3. a = 32.0, b = 22.6

4. c = 350, $\beta = 72.5°$

5. b = 17.8, c = 35.6

Find the acute angle ϴ (to nearest tenth degree) using tables or a
  calculator.

6. sin ϴ = 0.5667   11. csc ϴ = 1.858

7. tan ϴ = 1.026   12. sin ϴ = 0.2358

8. cos ϴ = 0.7354   13. cot ϴ = 1.442

9. sec ϴ = 1.506   14. cos ϴ = 0.3722

10. sin ϴ = 0.7967   15. tan ϴ = 5.782

16. An observer from a tower that is 200 ft. above sea level notices that
     the angle of depression to a boat is 15.2°. Assume that the base of
     the tower is at sea level. How far is the boat from the base of the
     tower?

17. The angle of elevation of the top of a building from a distance of
     1000 ft. is 25.3°. How high is the building? Neglect the height of
     the surveying instrument.

18. From a spot on one side of a river directly opposite a rock on the
     other side of the river a surveyor walks 110 ft. along the edge of
     the river. He then measures the angle between his line of sight to
     the rock and the spot directly opposite the rock. He finds it to be
     56.7°. What is the width of the river?

19. The angle of depression from the top of a tower to a boat that is 2000 ft. from the base of the tower is 23.2°. How high is the tower?

20. The angle of depression from the top of a 942 ft. tower with respect to a car that is at the same level as the base of the tower is 32.5°. How far from the base of the tower is the car?

21. Determine the height of a tree that casts a shadow of 42.0 ft. when the angle of elevation of the sun is 50.0°.

22. A guy wire to the top of a telephone pole that is 50.6 feet high is staked to the ground at a distance of 72.6 feet from the base of the pole. Find the angle between the wire and the ground.

23. From a tower, 572 feet above level ground, the angle of depression of the base of a building is 27.5°. Find the distance from the base of the tower to the base of the building.

8.3 Functions of Angles.

In order to deal with angles that are larger than $90°$, a more general definition for the trigonometric functions of angles is needed.

8.3.1 **Definition.** If the point (x,y) is on the terminal side of angle Θ, and r is the distance from the origin to the point (x,y).

$$\sin \Theta = \frac{y}{r} \qquad\qquad \csc \Theta = \frac{r}{y}$$

$$\cos \Theta = \frac{x}{r} \qquad\qquad \sec \Theta = \frac{r}{x}$$

$$\tan \Theta = \frac{y}{x} \qquad\qquad \cot \Theta = \frac{x}{y}$$

Note that r is always positive. The trigonometric functions will be positive, negative or zero as determined by the values of x and y.

If an angle is in standard position in the plane, the values of the trigonometric functions for that angle are determined by any point (x,y) on its terminal side. The distance, r, can be found by using the Pythagorean Theorem, i.e., $r = \sqrt{x^2 + y^2}$.

8.3.2 **Definition.** The reference angle, α, of any angle, Θ, in standard position that is not a quadrantal angle is the acute angle formed by the terminal side of Θ and the x-axis.

The values of the coordinates of a point (x,y) in each quadrant are shown in the following diagram.

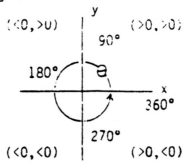

(<0,>0)   y   (>0,>0)
          90°
180°   θ
          x
          360°
          270°
(<0,<0)   (>0,<0)

If the definitions are applied in obtaining the six trigonometric functions, the ratios obtained for:

(a) $0° < θ < 90°$ are all positive values;

(b) $90° < θ < 180°$ only the sine and cosecant function are positive;

(c) $180° < θ < 270°$ only the tangent and the cotangent functions are positive;

(d) $270° < θ < 360°$ only the cosine and secant functions are positive.

8.3.3 Example. Determine the reference angle, $α$, for the given $θ$.

(a)

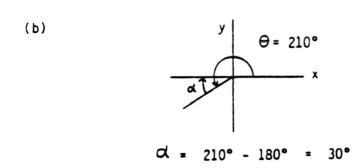

y   θ = 150°
α   x

$α = 180° - 150° = 30°$

(b)

y   θ = 210°
    x
α

$α = 210° - 180° = 30°$

332

(c)

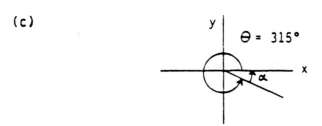

$$\alpha = 360° - 315° = 45°$$

(d)

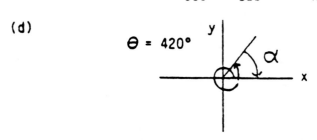

$$\alpha = 420° - 360° = 60°$$

<u>8.3.4 Example</u>. Given the point (-4,3) on the terminal side of as shown in the diagram below. Determine the six trigonometric functions of

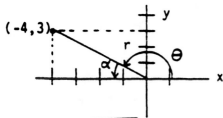

<u>Solution</u>: $r = \sqrt{(-4)^2 + (3)^2} = 5$

Applying definition 8.3.1, we obtain:

$\sin \theta = \dfrac{3}{5}$ $\csc \theta = \dfrac{5}{3}$

$\cos \theta = \dfrac{-4}{5}$ $\sec \theta = \dfrac{-5}{4}$

$\tan \theta = \dfrac{-3}{4}$ $\cot \theta = \dfrac{-4}{3}$

In the above example, note that the values of the trigonometric functions of the reference angle, α , have the same absolute values as the corresponding function values of θ .

The sides of the reference triangle, ABC, can be used in evaluating the trigonometric functions of α. By attaching the proper sign to a trigonometric function of α according to the quadrant in which the terminal side lies, one obtains the corresponding trigonometric function of θ.

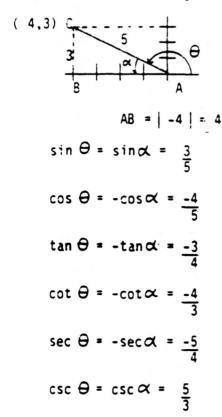

$$AB = |-4| = 4$$

$$\sin \theta = \sin \alpha = \frac{3}{5}$$

$$\cos \theta = -\cos \alpha = \frac{-4}{5}$$

$$\tan \theta = -\tan \alpha = \frac{-3}{4}$$

$$\cot \theta = -\cot \alpha = \frac{-4}{3}$$

$$\sec \theta = -\sec \alpha = \frac{-5}{4}$$

$$\csc \theta = \csc \alpha = \frac{5}{3}$$

<u>SELF QUIZ</u>. Determine the trigonometric functions of θ if a point on the terminal side has the coordinates (-1,1) by using the reference angle together with the reference triangle.

ANS. $\sin \theta = \sin \alpha = \frac{1}{\sqrt{2}}$

$$\cos \theta = -\cos \alpha = \frac{-1}{\sqrt{2}}$$

$$\tan \theta = -\tan \alpha = -1$$

$$\cot \theta = -\cot \alpha = -1$$

$$\sec \theta = -\sec \alpha = -\sqrt{2}$$

$$\csc \theta = \csc \alpha = \sqrt{2}$$

8.3.5 Example. Determine the values of the given trigonometric functions by use of tables or the use of a calculator.

(a) sin 150°, (b) sin 230°, (c) tan 230°, (d) cos 260°,

(e) tan 320°, (f) sec 165°, (g) csc 250°.

Solution: (a) sin 150° = sin 30° = 0.5000

(b) sin 230° = -sin 50° = -0.7660

(c) tan 230° = tan 50° = 1.1918

(d) cos 260° = -cos 80° = -0.1736

(e) tan 320° = -tan 40° = -0.8391

(f) sec 165° = -sec 15° = -1.0353

(g) csc 250° = -csc 70° = -1.0642

Thus far only positive angles have been considered. One can also consider negative angles.

8.3.6 Example.

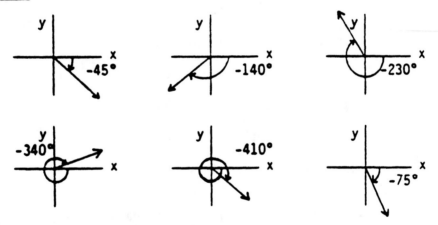

Regardless of the direction, trigonometric functions of Θ are still determined by the reference angle.

8.3.7 Example. Determine the trigonometric functions of $\Theta = -140°$. Express as trigonometric functions of the refernce angle before determining numerical values.

Solution: Answers are to four significant digits.

$$\sin(-140°) = -\sin 40° = -0.6428$$

$$\cos(-140°) = -\cos 40° = -0.7660$$

$$\tan(-140°) = \tan 40° = 0.8391$$

$$\cot(-140°) = \cot 40° = 1.192$$

$$\sec(-140°) = -\sec 40° = -1.305$$

$$\csc(-140°) = -\csc 40° = -1.556$$

SELF QUIZ. Determine the trigonometric functions of $\Theta = -230°$. Express as trigonometric functions of the reference angle before determining the numerical value. Answers should be to four significant digits.

ANS. $\sin \Theta = \sin 50° = 0.7660$

$\cos \Theta = -\cos 50° = -0.6428$

$\tan \Theta = -\tan 50° = -1.192$

$\cot \Theta = -\cot 50° = -0.8391$

$\sec \Theta = -\sec 50° = -1.556$

$\csc \Theta = \csc 50° = 1.305$

8.3 Exercise Set. All Θ are given in standard position. Find the value of the reference angle, α, of the given Θ.

1. 215°

2. 173°

3. 325°

4. 25°

5. -205°

6. -175

Express the given trigonometric functions in terms of the same function of a positive acute angle.

7. cos 115°

8. sin 225°

9. tan 130°

10. tan 230°

11. cos (-155°)

12. cot (-235°)

13. sin (-27°)

14. csc 217°

15. tan 715°

16. sec (-235°)

17. cot (-254°)

18. cot 700°

Given: the terminal side of Θ, in standard position, passes through the given point. Draw a graph and make use of the reference triangle in finding the six trigonometric functions of Θ. Write answers to four significant digits.

19. (-3,-12)

20. (8,-6)

21. (-5,12)

22. (9,-15)

23. (-4,-3)

24. (-15,8)

25. (-7,-24)

26. (-5,-12)

27. (-6,-8)

28. (-12,6)

29. (-3,-1)   30. (4,-5)

Find all $0° < \theta < 360°$ to nearest tenth degree.

31. $\sin \theta = 0.5862$ 34. $\sec \theta = -1.348$

32. $\cos \theta = -0.7632$ 35. $\csc \theta = -1.577$

33. $\tan \theta = 2.372$ 36. $\cot \theta = -2.786$

Under the given conditions, determine the quadrant in which θ must have its terminal side.

37. $\sin \theta < 0$ and $\cos \theta > 0$ 40. $\tan \theta < 0$ and $\cos \theta < 0$

38. $\cos \theta < 0$ and $\tan \theta > 0$ 41. $\cos \theta > 0$ and $\tan \theta < 0$

39. $\sin \theta < 0$ and $\tan \theta > 0$ 42. $\tan \theta < 0$ and $\sin \theta > 0$

Determine the two possible quadrants in which θ has its terminal side.

43. $\cos \theta < 0$ 46. $\tan \theta < 0$

44. $\tan \theta > 0$ 47. $\sec \theta < 0$

45. $\cos \theta > 0$ 48. $\cot \theta > 0$

Chapter 8

8.1 1) a

b

c

d

	sin	cos	tan	cot	sec	csc
2) a	0.8000	0.6000	1.333			
b	0.9231	0.3846	2.400			
c	0.6988	0.7155	0.9767			
d	0.7372	0.6757	1.091			
3) α	0.5000	0.8660	0.5774	1.732	1.155	2.000
β	0.8660	0.5000	1.732	0.5774	2.000	1.115
4) α	0.7071	0.7071	1.000	1.000	1.414	1.414
β	0.7071	0.7071	1.000	1.000	1.414	1.414
5) α	0.3846	0.9231	0.4167	2.400	1.083	2.600
β	0.9231	0.3846	2.400	0.4167	2.600	1.083
6) α	0.4706	0.8824	0.5333	1.875	1.133	2.125
β	0.8824	0.4706	1.875	0.5333	2.125	1.133
7) α	0.2800	0.9600	0.2917	3.429	1.042	3.571
β	0.9600	0.2800	3.429	0.2917	3.571	1.042

	sin	cos	tan	cot	sec	csc
8) α	0.6000	0.8000	0.7500	1.333	1.250	1.667
β	0.8000	0.6000	1.333	0.7500	1.667	1.250
9) α	0.6000	0.8000	0.7500	1.333	1.250	1.667
β	0.8000	0.6000	1.333	0.7500	1.667	1.250
10) α	0.9600	0.2800	3.429	2.917	3.571	1.042
β	0.2800	0.9600	2.917	3.429	1.042	3.571
11) α	0.8824	0.4706	1.875	0.5333	2.125	1.133
β	0.4706	0.8824	0.5333	1.875	1.133	2.125
12) α	0.7454	0.6667	1.118	0.8944	1.500	1.342
β	0.6667	0.7454	0.8944	1.118	1.342	1.500
13) α	0.9600	0.2800	3.429	0.2917	2.571	1.042
β	0.2800	0.9600	0.2917	3.429	1.042	2.571
14) α	0.9231	0.3846	2.400	0.4167	2.600	1.083
β	0.3846	0.9231	0.4167	2.400	1.083	2.600
15) θ	0.8000	0.6000	1.333	0.7500	1.667	1.250
16) θ	0.3846	0.9231	0.4167	2.400	1.083	2.500
17) θ	0.9231	0.3846	2.400	0.4167	2.600	1.083
18) θ	0.8660	0.5000	1.732	0.5774	2.000	1.155
19) θ	0.8000	0.6000	1.333	0.7500	1.667	1.250
20) θ	0.9798	0.2000	4.899	0.2041	5.000	1.021
21) θ	0.9600	0.2800	3.429	0.2917	3.571	1.042
22) θ	0.8660	0.5000	1.732	0.5774	2.000	1.155
23) θ	0.2500	0.9682	0.2582	3.873	1.033	4.000
24) θ	0.9285	0.3714	2.500	0.4000	2.693	1.077
25) θ	0.6000	0.8000	0.7500	1.333	1.250	1.667
26) θ	0.9600	0.2800	3.429	0.2917	3.571	1.042

	sin	cos	tan	cot	sec	csc
27) θ	0.8824	0.4706	1.875	0.5333	2.125	1.133
28) θ	0.9040	0.4276	2.114	0.4730	2.339	1.106
29) θ	0.7371	0.6757	1.091	0.9167	1.480	1.357
30) θ	0.3590	0.9333	0.3846	2.600	1.071	2.786

31) cos 40°   32) csc 55°   33) sin 50°

34) tan 65°   35) cot 33°   36) sec 15°

8.2 1) β = 49.8°, a = 9.68, b = 11.5

2) β = 54.5°, c = 15.2, a = 8.92

3) c = 39.2, α = 54.8°, β = 35.2°

4) α = 17.5°, a = 105, b = 334

5) a = 30.8, α = 60°, β = 30°

6) 34.5°   7) 45.7°   8) 42.7°

9) 48.4°   10) 52.8°   11) 32.6°

12) 13.6°   13) 34.7°   14) 68.1°

15) 80.2°   16) 736 ft.   17) 473 ft.

18) 167 ft.   19) 857 ft.   20) 1,479 ft.

21) 50.1 ft.   22) 34.9°   23) 1,099 ft.

8.3 1) 35°   2) 7°   3) 35°

4) 25°   5) 25°   6) 5°

7) -cos 65°   8) -sin 45°   9) -tan 50°

10) tan 50°   11) -cos 25°   12) -cot 55°

13) -sin 27°   14) -sin 37°   15) -tan 5°

16) -sec 55°   17) -cot 74°   18) -cot 20°

		sin	cos	tan	cot	sec	csc
19)	θ	-0.9677	-0.2419	4.000	0.2500	-4.133	-1.033
20)	θ	-0.6000	0.8000	-0.7500	-1.333	1.250	-1.667
21)	θ	-0.9231	-0.3846	2.400	0.4167	-2.600	-1.083
22)	θ	-0.8824	0.4706	-1.875	-0.5333	2.125	-1.133
23)	θ	-6.000	-0.8000	0.7500	1.333	-1.250	-1.667
24)	θ	0.4706	-0.8824	-0.5333	-1.875	-1.133	2.125
25)	θ	-0.9600	-0.2800	3.429	0.2917	-3.571	-1.042
26)	θ	-0.9231	-0.3846	2.400	0.4167	-2.600	-1.083
27)	θ	-0.8000	-0.6000	1.333	0.7500	-1.667	-1.250
28)	θ	0.4478	-0.8955	-0.5000	-2.000	-1.117	2.233
29)	θ	-0.5000	-0.8660	0.5774	1.732	-1.155	-2.000
30)	θ	-0.7809	0.6247	-1.250	-0.8000	1.601	-1.281

31) {32.4°, 147.6°}   32) {139.7°, 220.3°}   33) {67.1°, 247.1°}

34) {137.9°, 222.1°}   35) {219.4°, 320.6°}   36) {160.3°, 340.3°}

37) IV   38) III   39) III

40) II   41) IV   42) II

43) II and III   44) I and III   45) I and IV

46) II and IV   47) II and III   48) I and III

8.4		sin	cos	tan	cot	sec	csc
1)	60°	$\dfrac{\sqrt{3}}{2}$	$\dfrac{1}{2}$	$\sqrt{3}$	$\dfrac{1}{\sqrt{3}}$	2	$\dfrac{2}{\sqrt{3}}$
2)	150°	$\dfrac{1}{2}$	$\dfrac{-\sqrt{3}}{2}$	$\dfrac{-1}{\sqrt{3}}$	$-\sqrt{3}$	$\dfrac{-2}{\sqrt{3}}$	2
3)	60°	$\dfrac{-\sqrt{3}}{2}$	$\dfrac{1}{2}$	$-\sqrt{3}$	$\dfrac{-1}{\sqrt{3}}$	2	$\dfrac{-2}{\sqrt{3}}$
4)	45°	$\dfrac{1}{\sqrt{2}}$	$\dfrac{1}{\sqrt{2}}$	1	1	$\sqrt{2}$	$\sqrt{2}$